ENGINEERING
METROLOGY

THE PERGAMON MATERIALS ENGINEERING PRACTICE SERIES

Editorial Board

Chairman: D. W. HOPKINS, University College of Swansea
J. R. BARRATT, British Steel Corporation
T. BELL, University of Birmingham
G. E. SHEWARD, UKAEA, Springfields Laboratories
A. J. SMITH
Secretary: A. POST

OTHER TITLES IN THE SERIES

ALLSOP & KENNEDY	Pressure Diecasting, Part 2
BYE	Portland Cement
DAVIES	Protection of Industrial Power Systems
HENLEY	Anodic Oxidation of Aluminium and Its Alloys
HOLLAND	Microcomputers for Process Control
LANSDOWN	Lubrication
LANSDOWN & PRICE	Materials to Resist Wear
MASKALL & WHITE	Vitreous Enamelling
MORGAN	Tinplate and Modern Canmaking Technology
NEMENYI	Controlled Atmospheres for Heat Treatment
PARRISH & HARPER	Production Gas Carburising
SHEWARD	High Temperature Brazing in Controlled Atmospheres
UPTON	Pressure Diecasting, Part 1
WILLIAMS	Troubleshooting on Microprocessor Based Systems

NOTICE TO READERS

Dear Reader

An Invitation to Publish in and Recommend the Placing of a Standing Order to Volumes Published in this Valuable Series.

If your library is not already a standing/continuation order customer to this series, may we recommend that you place a standing/continuation order to receive immediately upon publication all new volumes. Should you find that these volumes no longer serve your needs, your order can be cancelled at any time without notice.

The Editors and the Publisher will be glad to receive suggestions or outlines of suitable titles, reviews or symposia for editorial consideration: if found acceptable, rapid publication is guaranteed.

ROBERT MAXWELL
Publisher

ENGINEERING METROLOGY

D. M. ANTHONY
Formerly Head of Product Assurance, Rolls Royce, Bristol, UK

PERGAMON PRESS

OXFORD · NEW YORK · BEIJING · FRANKFURT
SÃO PAULO · SYDNEY · TOKYO · TORONTO

U.K.	Pergamon Press, Headington Hill Hall, Oxford OX3 0BW, England
U.S.A.	Pergamon Press, Maxwell House, Fairview Park, Elmsford, New York 10523, U.S.A.
PEOPLE'S REPUBLIC OF CHINA	Pergamon Press, Qianmen Hotel, Beijing, People's Republic of China
FEDERAL REPUBLIC OF GERMANY	Pergamon Press, Hammerweg 6, D-6242 Kronberg, Federal Republic of Germany
BRAZIL	Pergamon Editora, Rua Eça de Queiros, 346, CEP 04011, São Paulo, Brazil
AUSTRALIA	Pergamon Press Australia, P.O. Box 544, Potts Point, N.S.W. 2011, Australia
JAPAN	Pergamon Press, 8th Floor, Matsuoka Central Building, 1-7-1 Nishishinjuku, Shinjuku-ku, Tokyo 160, Japan
CANADA	Pergamon Press Canada, Suite 104, 150 Consumers Road, Willowdale, Ontario M2J 1P9, Canada

Copyright © 1986 Pergamon Books Ltd.

All Rights Reserved. No part of this publication may be reproduced, stored in a retrieval system or transmitted in any form or by any means: electronic, electrostatic, magnetic tape, mechanical, photocopying, recording or otherwise, without permission in writing from the publishers.

First edition 1986

Library of Congress Cataloging in Publication Data
Anthony, D. M.
Engineering metrology
(The Pergamon materials engineering practice series)
Includes index.
1. Mensuration. I. Title. II. Series.
T50.A57   1986   602.8'7   86–9438

British Library Cataloguing in Publication Data
Anthony, D. M.
Engineering metrology.—(The Pergamon materials engineering practice series)
1. Mensuration
I. Title
620'.0044   T50

ISBN 0-08-028682-8 Hard cover
ISBN 0-08-028683-6 Flexicover

Printed in Great Britain by A. Wheaton & Co. Ltd., Exeter

Materials Engineering Practice

FOREWORD

The title of this series of books "Materials Engineering Practice" is well chosen since it brings to our attention that in an era where science, technology and engineering condition our material standards of living, the effectiveness of practical skills in translating concepts and designs from the imagination or drawing board to commercial reality, is the ultimate test by which an industrial economy succeeds.

The economic wealth of this country is based principally upon the transformation and manipulation of *materials* through *engineering practice*. Every material, metals and their alloys and the vast range of ceramics and polymers has characteristics which require specialist knowledge to get the best out of them in practice, and this series is intended to offer a distillation of the best practices based on increasing understanding of the subtleties of material properties and behaviour and on improving experience internationally. Thus the series covers or will cover such diverse areas of practical interest as surface treatments, joining methods, process practices, inspection techniques and many other features concerned with materials engineering.

It is to be hoped that the reader will use this book as the base on which to develop his own excellence and perhaps his own practices as a result of his experience and that these personal developments will find their way into later editions for future readers. In past years it may well have been true that if a man made a better mousetrap the world would beat a path to his door. Today, however, to make a better mousetrap requires more direct communication between those who know how to make the better mousetrap and those who wish to know. Hopefully this series will make its contribution towards improving these exchanges.

MONTY FINNISTON

Acknowledgements

I am indebted to the following firms who have provided material and illustrations:
 J. E. Baty & Co. Ltd.
 BM Metrology Services
 British Standards Institution
 BSG (Machinery Sales) Ltd.
 Crown Windley
 Dixi & Associates
 OMT-Optical Measuring Tools
 Eimeldingen (GB) Ltd.
 Federal Gauges Ltd.
 Ferranti Metrology Systems
 Findlay, Irvine Ltd.
 Gleason Works Ltd.
 Hahn & Kolb (GB) Ltd.
 Horstmann-Coventry Gauge Ltd.
 C. E. Johansson Ltd.
 Kemco Measuring Systems
 KeyMed Industrial
 George Kuikka Ltd.
 Kroplin (Great Britain) Ltd.
 E. Leitz (Instruments) Ltd.
 LK Tool Co. Ltd.
 Magnaflux Ltd.
 Machsize Ltd.
 Marposs Electronic Gauges Ltd.
 Mitutoyo (UK) Ltd.
 Neill Tools Ltd.
 Newall Electronics
 Macro CNC Machine Tools Ltd.
 Rank Taylor Hobson Ltd.
 Renishaw Metrology Ltd.

Rolls-Royce Ltd.
Rubert & Co.
Scantron Ltd.
Select Gauges Ltd.
Sigma Ltd.
SIP-Dixi (UK) Ltd.
Stanhope Machine Tools Ltd.
Temco Tools Ltd.
Tesa Metrology Ltd.
TGM Metrology
Trimos-Sylvac Metrology Ltd.
Vernon Gauging Systems Ltd.
Verdict Gauges (Sales) Ltd.

I am particularly grateful for the co-operation of Rolls-Royce Ltd., who were my employers for most of the period during which the book was written, and for the help given to me by many colleagues.

Last but not least I must express my gratitude to editor Don Hopkins for his invaluable encouragement and advice.

DEREK M. ANTHONY
February 1986

Contents

Introduction   xiii

1. HISTORY   1
   - 1.1   The Yard and the Metre   1
   - 1.2   Millionth Measuring Instruments   2
   - 1.3   Optical Length Standards   3
   - 1.4   Industrial Measurement   4

2. PHILOSOPHY OF MEASUREMENT   7
   - 2.1   Process Performance   7
   - 2.2   Specification Validity   8
   - 2.3   Measurement Principles   9
   - 2.4   Selection of Equipment   10
   - 2.5   Quantity   11
   - 2.6   Accuracy   12
   - 2.7   Distortion   13
   - 2.8   Parallax   14
   - 2.9   General Measuring Accuracy   15
   - 2.10   Use of Comparison   15

3. SPECIFYING THE REQUIREMENT   17
   - 3.1   Tolerance Principles   17
   - 3.2   Standard Tolerance Systems   19
   - 3.3   Geometrical Tolerances   24
   - 3.4   Process Capability   27

4. BASIC MECHANICAL MEASURING INSTRUMENTS   29
   - 4.1   The Steel Rule   29
   - 4.2   Vernier Scales   30
   - 4.3   Vernier Caliper, Height Gauge and Depth Gauge   31
   - 4.4   Surface Tables, Box Cubes and Vee Blocks   33
   - 4.5   Display Systems   37

	4.6	The Dial Indicator	39
	4.7	Electronic and Pneumatic Indicators	43
	4.8	The Micrometer	46
	4.9	Bore Gauges, Plug Gauges	52
	4.10	Ring Gauges, Snap Gauges	55

5. **LINEAR MEASUREMENTS AND MULTI-AXIS MEASURING MACHINES** — 58
    - 5.1 Electronic Measuring Systems — 58
    - 5.2 Multi-axis Measuring Machines — 60
    - 5.3 Measuring Probes — 68

6. **ANGULAR MEASUREMENT** — 74
    - 6.1 Specification of Angles — 74
    - 6.2 Use of Linear Conversion — 76
    - 6.3 The Protractor — 76
    - 6.4 The Spirit Level — 79
    - 6.5 The Sine Bar — 81
    - 6.6 Rotating Tables — 83
    - 6.7 Roundness — 89

7. **FLATNESS AND SURFACE FINISH** — 91
    - 7.1 Quantification of Surface Finish — 91
    - 7.2 Surface Finish Units — 94
    - 7.3 Measurement of Surface Finish — 95
    - 7.4 Surface Finish Measuring Machines — 96
    - 7.5 Use of Replicas — 100
    - 7.6 Flatness — 101

8. **GEAR TEETH** — 102
    - 8.1 Types of Gears — 102
    - 8.2 The Involute Form — 103
    - 8.3 Involute Measurement — 104
    - 8.4 Tooth Thickness and Pitch — 110
    - 8.5 Rolling Tests — 115
    - 8.6 Tooth Marking Tests — 119

9. **MEASUREMENT OF CONTOURED SURFACES** — 121
    - 9.1 Optical Projection — 121
    - 9.2 Templates — 127
    - 9.3 Multi-probe Measurement — 128

10. **SCREW THREADS** — 132
    - 10.1 Definition of a Screw Thread — 132
    - 10.2 Standard Threads — 134
    - 10.3 Thread Measurement — 135

	10.4	Virtual Effective Diameter	141
	10.5	Internal Threads	141
	10.6	Measurement of Production Threads	142
	10.7	Plug Gauges	142
	10.8	Ring Gauges	143
	10.9	Thread Tolerances	144
	10.10	Coating of Threads	145
	10.11	Gauge Tolerances	146
	10.12	Other Thread Measuring Methods	147
11.	**VIEWING DEVICES AND OPTICAL MEASUREMENT**		149
	11.1	Visual Assessment	149
	11.2	Projection	151
	11.3	Non-contact Measurement	155
	11.4	Telescopes and Collimators	158
	11.5	The Autocollimator	162
	11.6	Interferometry	165
	11.7	Laser	170
12.	**ELECTRONICS, COMPUTERS AND OTHER METROLOGY TECHNIQUES**		174
	12.1	Computers	174
	12.2	Function Control	174
	12.3	Output Processing	176
	12.4	Computer Types	178
	12.5	Non-destructive Testing	183
	12.6	Radiography	183
	12.7	Ultrasonics	185
	12.8	Gamma Radiation	186
	12.9	Eddy Current Methods	186
13.	**STANDARDS ROOM, LABORATORY AND SPECIAL PURPOSE MEASUREMENT**		188
	13.1	Objectives	188
	13.2	Instrument Control Procedure	188
	13.3	Accuracy Requirements	189
	13.4	Effect of Temperature	190
	13.5	Effect of Distortion	190
	13.6	Basic Requirements	191
	13.7	Gauge Blocks	191
	13.8	End Bars	196
	13.9	Toolmaker's Microscope	196
	13.10	Special Purpose Measuring Machines	201

14.	IN-PROCESS AND IN-CYCLE MEASUREMENT	204
	14.1 In-process Gauging	204
	14.2 In-cycle Gauging	208 |//
Postscript		217
Index		219

Introduction

The semantics of the quality vocabulary are today such that the word "inspection" is somewhat old-fashioned, and there is even an inclination to dismiss the activity as unnecessary.

It is, however, an absolutely essential tool for satisfying the quality assurance requirements of any engineering product, since their primary constituent will certainly be an assurance that the details of the product conform to its design specification.

It must be the objective of any engineering enterprise to market a product with a level of performance and reliability which will at least satisfy the customer. The degree to which this is achieved will establish the reputation of the enterprise and make a large contribution to its continued profitable business.

The design specification of the product will have been subjected to a process of development, aimed at achieving the necessary level, and it is the responsibility of the quality assurance function to be satisfied that all examples of the product conform in every detail to the design requirement.

The design will determine the detail of the product by means of a large number and variety of parameters, and inspection is the only positive way of ensuring that the product conforms to the requirement in every detail.

This book aims to give guidance on the principles of inspection, through the science of metrology, and to indicate the options available for carrying out the task.

It covers methods currently available, and expected to be available in the foreseeable future, for all forms of engineering measurement, and, as such, it provides a basic textbook for those who wish to study engineering metrology.

Perhaps more important, it also offers guidance, both to the individual and the organisation, on the selection of the most efficient and economical means of carrying out a measuring task. In this respect it is hoped that it will be particularly useful to the small company which does not necessarily have, or wish to use, expensive resources for investigation or development of metrology methods.

Whilst the book covers many of the more elaborate, and inevitably expensive, techniques, it endeavours to illustrate how much can be achieved using simple and comparatively cheap equipment.

There will be many situations where the need for a high level of accuracy will involve high cost of both production and inspection, but, even in these cases, the book may assist both in avoiding over specification of dimensional accuracy, and in encouraging the use of the most economical inspection methods.

Chapter 1

History

From the earliest times, man found that he had a need to be able to measure. It is likely that he survived for many centuries requiring only to measure length, and his obvious means of doing this was by comparison with parts of his body. The Imperial foot, of course, has its origin in this "system", and it is only in recent years that it has become obsolescent with the universal introduction of the metric system.

Otherwise, units of measurement were determined throughout the world on a very local basis, and the development of trading between nations created the need for measurement standards which were internationally recognised.

It would seem that a great deal of work on this subject was done in the seventeenth century, and, indeed, the proposal to use one ten-millionth of a quadrant of the earth as the standard length unit was first made around the middle of this century. It was, however, more than a century later, in 1971, that the French government agreed to its adoption, specifically, as one ten-millionth of a quadrant from North Pole to Equator.

1.1 THE YARD AND THE METRE

Probably the earliest standard of any precision was a bronze bar made in 1760 which was adopted as the British standard of length in 1824. This standard was destroyed by fire 10 years later and a replacement was made as nearly as possible to the same length. The geometry of this replacement was more scientific in that the measurement was between two fine lines, scribed on gold plugs let into the bar so that their surfaces were at the centre of the section. This bar remained the "official" Imperial Standard Yard until 1963.

The metre standard was produced in 1799 as rectangular section platinum end bar, and it was identified as the "Metre des Archives". In 1870 an international conference on length held in Paris decided

2 History

to produce a new metre prototype. This was made from platinum-iridium alloy and had a cross-section and dimensions as shown in Fig. 1.1. Markings were scribed on the 4-mm wide plateau on the centre line.

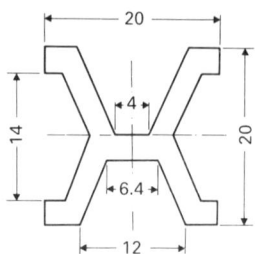

FIGURE 1.1

Thirty-two prototypes were made, and in 1889 No. 6 was decreed to be the closest to the "Metre des Archives" and adopted as the international metre at the first International Conference of Weights and Measures held in that year.

In the United States the yard standard was adopted in 1830, and in 1866 the use of the metric system was also approved, and the metre was established as 39.37 in. In 1895, however, the British determined the value as 39.370113 in. This difference, although appearing to be small, was the cause of many metrology problems between Britain and the United States. Further comparisons were made in 1922, 1932 and 1947, all of which gave an increasingly larger figure to the Imperial value of the metre, and this led to the conclusion that the Imperial Standard Yard was steadily shrinking. In order to make the situation compatible, Britain and the United States agreed in 1963 that the relationship should be that 1 in exactly equalled 25.4 mm.

1.2 MILLIONTH MEASURING INSTRUMENTS

In parallel with the establishment of internationally agreed length standards, development was necessarily proceeding on means of measuring to the degree of accuracy required to make such standards useful.

The most ingenious devices appeared during the period between the wars. Two of these are shown in Figs. 1.2 and 1.3.

Figure 1.2 is the Brooks level comparator. This is basically a refined form of spirit level and compares two end bars by standing

Optical Length Standards 3

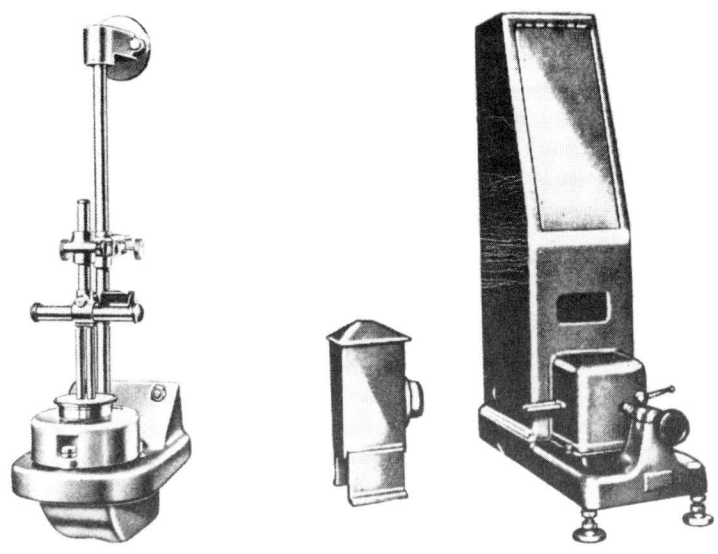

FIGURE 1.2   FIGURE 1.3

them vertically on a flat rotating table and placing the feet of the level on their upper ends. The position of the bars is then reversed by rotating the table and a second level reading taken. The difference between the readings will then be double the differences in length of the end bars.

The instrument in Fig. 1.3 is the Eden–Rolt comparator which uses an ingenious combination of mechanical and optical magnification. Both of these instruments are capable of resolution to an accuracy of one-millionth of an inch.

1.3 OPTICAL LENGTH STANDARDS

At a surprisingly early stage, the idea of determining length standards in terms of the wavelength of light was considered, and in the late nineteenth century the metre could be quoted as a number of wavelengths of cadmium light. Wartime developments in atomic physics disclosed that radiations of certain isotopes were more nearly monochromatic than the natural elements, and the General Conference on Weights and Measures held in 1960 adopted the proposal that the metre should be defined as 1650763.73 wavelengths of the light emitted by krypton-86 *in vacuo*.

1.4 INDUSTRIAL MEASUREMENT

As far as general measurement is concerned, industry was a long way behind the laboratories. Even in the late nineteenth century, fractional measurements to an accuracy of 1/32nd of an inch were considered to be high precision. Early devices which were sufficiently advanced to read in decimals were unsuccessful because such a level of accuracy was thought to be quite unnecessary.

Perhaps the reason for this is that engineering products requiring precision were made in very small quantities, and it was normal for mating parts requiring a close fit to be made to suit each other by means of hand work at the assembly stage. The First World War, however, created a situation in which large quantities of precision armaments were required. This, in turn, demanded close-fitting parts which could be assembled together, and were internationally interchangeable, without the need for hand work.

The development of the aircraft industry, in peace and war, is also likely to have had a strong influence in demanding a new order of magnitude in reliability and interchangeability. An important factor in achieving these qualities is the consistency with which one example from a large quantity of a product is the same as another in all its significant parameters. This requires a disciplined system of control over manufacturing variations, a requirement which was satisfied by the introduction of quantified manufacturing tolerances.

In order to ensure that specified tolerances were being maintained, it then became necessary to have much more accurate production measuring equipment than had hitherto been available.

Some of the earliest precision machines for production measurement were aimed at screw threads, and this led to the development of the micrometer. Several instruments using the principle of a screw thread to measure length were made during the nineteenth century, and hand micrometers, as shown in Fig. 1.4, which closely resemble those available today, appeared towards the end.

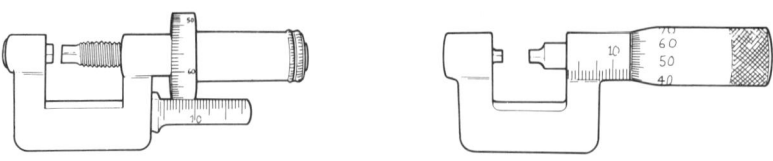

FIGURE 1.4

When once accepted, the micrometer rapidly became the most common precision length measuring device, and was in very wide use early in the twentieth century.

The other development which perhaps ranks with the micrometer in producing a step change in industrial measuring accuracy was the invention of gauge blocks. These were produced around 1900 by C. E. Johansson, then an inspector in a Swedish arms factory. He made his early gauges personally by lapping the end faces to the necessary high finish on an adapted treadle sewing machine. Present day sets of gauge blocks differ very little from those produced by Johansson at this time. They became the means throughout the world of establishing a local base for high precision measurement.

During the first half of the present century there was little in the way of fundamental change to the design of measuring equipment. Almost without exception, instruments were either mechanical or optical, and improvements were aimed at increasing accuracy by meticulous attention to detail.

The only machines available which might in any way be described as general purpose multi-axis were produced for standards room and metrology laboratory use, as the example in Fig. 1.5.

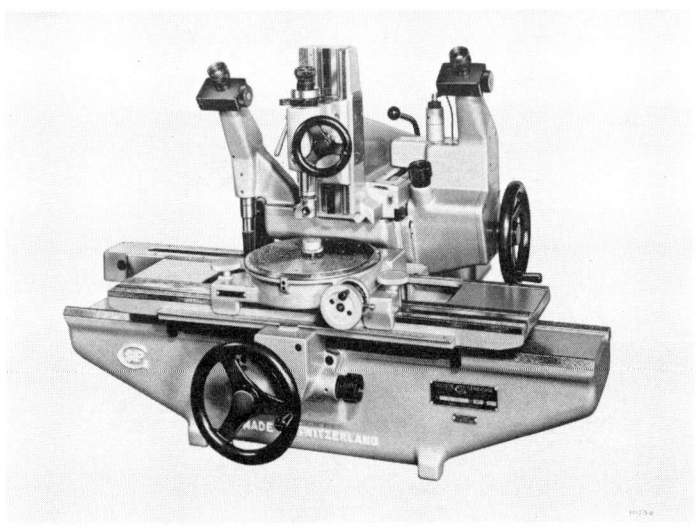

FIGURE 1.5

They were inevitably very expensive and capable of a much higher order of accuracy than necessary for the routine examination of production parts. In spite of this, such machines were found to be

6 History

valuable, certainly for production troubleshooting, and not infrequently for production measurement of small batches.

This pointed to a great demand for a multi-axis measuring machine of production standard accuracy and production equipment cost. A step change occurred in the 1950s with a quite sudden surge in the development of solid state electronics, and this opened the way for satisfying the need for a production general purpose measuring machine. When such a machine did arrive, and an early example is shown in Fig. 1.6, it came from the electronics industry rather than the traditional precision mechanical background.

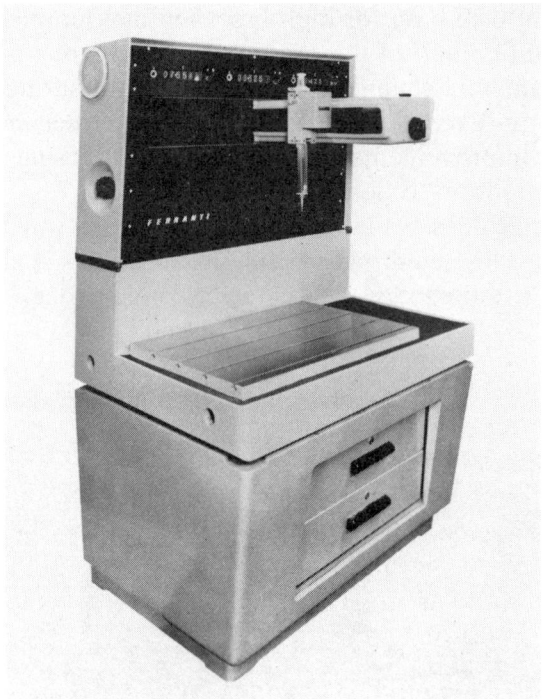

FIGURE 1.6

This electronic technology has since been very successfully combined with the use of non-traditional materials, particularly plastics and granite, to produce versatile, accurate and economical measuring machines such as those discussed in Chapter 5.

Most recent developments have been to take advantage of the electrical output of modern machines and to combine this with computer technology. Future trends are likely to expand this concept into measuring during the manufacturing cycle with automatic feedback to make corrective changes.

Chapter 2

Philosophy of Measurement

The basic objective of measuring is to obtain assurance that the article being measured, whether a piece of raw material, a completed part, an assembly or a finished product, is in accordance with a specification which has been offered to, or presented by, the customer. This concept is the use of measurement to achieve quality control.

In addition, however, awareness of actual sizes to an adequate degree of precision gives a feedback on the performance of the combination of man and machine, and this enables adjustments to be made so that they can perform at maximum efficiency.

Thus, whilst satisfying the specification by the finished product might be the basic objective, advantage can be taken of dimensional information which can readily be obtained from most of the metrology equipment currently available. This will enable an assessment of the consistency and efficiency of the manufacturing process to be made, and such information can also be used to assess the validity of the specification itself. This concept may be defined as the use of measurement to achieve quality assurance.

Thus the objective of measurement may be summarised as:

— to ensure that products supplied to the customer are within the agreed specification.
— to ensure, by monitoring process performance, that the number of unacceptable finished parts is as small as is economically practicable.
— to ensure, by comparing customer requirements and process complexity, that the specification satisfies the customer in the most efficient way.

2.1 PROCESS PERFORMANCE

The most important parameter available for measuring process performance is the proportion of unacceptable parts produced. This simple statistic, however, is not in itself completely satisfactory and it can, in fact, be misleading.

Firstly, it is necessary to divide the unacceptable parts into those which can, and those which cannot, be corrected. Even parts which are not correctable can often be either accepted under a concession for the defective feature, or salvaged by carrying out an additional process.

The philosophy of acceptance of a defective feature under concession is complex. It might, at first sight, seem logical to assume that if a dimension is acceptable even though outside its tolerance, then the tolerance should be widened. Whilst this is sometimes the case, generally acceptance is based upon considering the dimension concerned along with associated dimensions and their position within the tolerance band. It might, for example, be possible to accept a shaft with an oversize bore diameter provided that the outside diameter is towards the large end of its tolerance.

Salvaging might be defined as a supplementary process which restores a dimension to within its specified tolerance. Generally this involves adding material, which can be fairly easily achieved by one of many plating and metal spraying processes at present available.

Assuming that the process is technically sound, a salvage decision is usually economically based. Expressed simply, this will depend on whether the cost of the salvage operation is greater than the cost of making another part. Since salvages are normally highly individual, they will often be judged as uneconomic, but it is an indication of the danger of a simple "good/scrap" statistic that emphasis on keeping scrap low can result in uneconomical salvages.

The economic argument must again be introduced when determining the tolerable proportion of scrapped parts. It will invariably be uneconomical to produce no scrap at all, but it is not easy to determine the economic level. In certain processes, such as the casting of small parts, the most efficient production philosophy may be to produce large quantities very quickly, and very cheaply, and then sort the good from the bad. This could result in a large proportion of scrap but with the possibility that it can readily be melted and recast.

2.2 SPECIFICATION VALIDITY

The objective in assessing the specification is to be satisfied that the requirements of the customer will be met consistently, but at the same time making sure that the most efficient use is being made of available manufacturing resources.

In the case of a manufacturing process, for example, if a particular feature has a tolerance of 0.5 mm and measurement showed that it is being controlled within a variation of only 0.05 mm, it might be justifiably concluded that the manufacturing process is unnecessarily extravagant. If, on the other hand, a feature has a tolerance of 0.1 mm and it is found that the normal range of sizes produced by a reasonable process is 0.4 mm, investigation might be prompted into the need for the specification to demand such a tight tolerance.

2.3 MEASUREMENT PRINCIPLES

The use of metrology to achieve quality assurance rather than quality control is a comparatively recent trend. Some years ago measurement aimed to look at completed parts and merely to sort good from bad, and measuring equipment was based largely on the "go" or "no-go" principle, which does no more than indicate whether the feature is within the permissable tolerance or not. This principle does not permit analysis of process or specification, since no indication is given of where the actual size of the feature lies within the tolerance band, nor are any variations in the feature between one part and another apparent. The method involves the use of equipment which usually consists of cast or fabricated special purpose frames in which the part to be measured is located, and the various features are checked by devices of the type shown in Fig. 2.1.

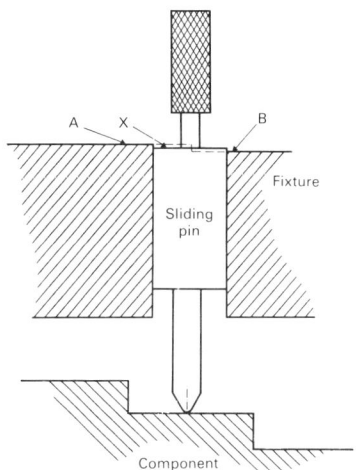

Figure 2.1

10 Philosophy of Measurement

In this case the dimension being checked is acceptable if the surface X is above surface A and below surface B. Finger touch is normally adequate to determine this, but, of course, a more accurate, and quantified result could be obtained if the check was done with a measuring instrument.

An example of a gauge which checks a number of dimensions on a large component in this way is shown in Fig. 2.2.

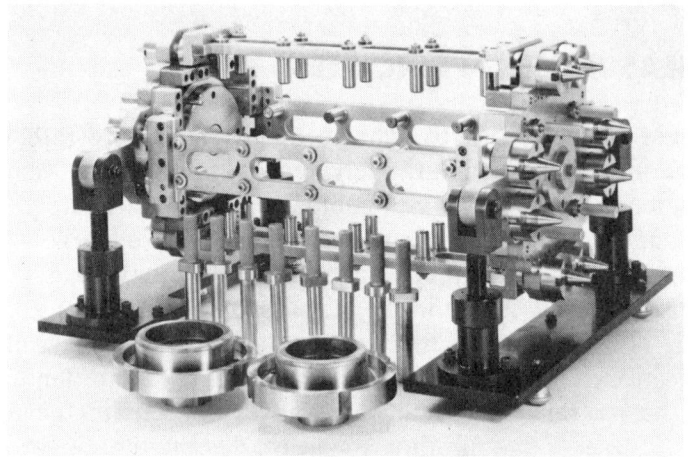

FIGURE 2.2

With the availability of electrical and electronic measuring probes and the multi-axis measuring machines, as described in Chapter 5, the "go/no-go" principle is rarely used by itself except in special circumstances. If simple "go/no-go" indications are required, they can be obtained easily from the output of electrical measuring probes. Modern measuring machines are basically capable of providing the quantitative information necessary for analysis, and, when their versatility is taken into account, they could well prove to be the most economical means of carrying out the majority of measuring tasks.

2.4 SELECTION OF EQUIPMENT

In determining the equipment, and method, to be used, several factors must be considered, the most important of which are the number of items to be measured, and the accuracy of measurement required.

2.5 QUANTITY

The sizes of both individual batches of parts, and the total quantities over a period are relevant, but whether the specific number is treated as large will depend on the type of product.

However, in any case, if the number of items is very small, say in single figures, nothing in the way of special equipment is likely to be justified, and it should be possible to obtain all the information necessary using the basic methods described in Chapter 4.

At the other end of the scale, the number of measurements must be considered large if, using basic equipment, they occupy a high proportion, say 10%, of available working time.

In these circumstances, a number of alternatives are available. A piece of equipment might be designed and manufactured to carry out one particular task, as the example in Fig. 2.3.

Figure 2.3

The other extreme might be a general multi-axis measuring machine, but this can be made temporarily "special purpose" by computer control using software designed for the specific task.

In between, there is a very useful choice of equipment consisting of modules which can be arranged on a universal base to produce another form of "temporary" special purpose equipment. An example of such equipment is shown in Fig. 2.4.

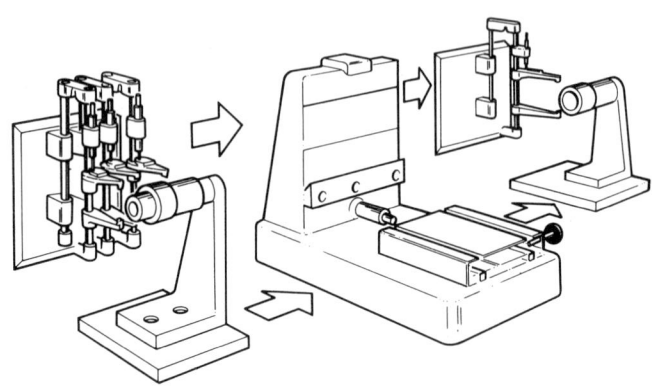

FIGURE 2.4

Equipment of this type is discussed further in Chapter 12.

The decision on which of the variety of alternatives is likely to be selected will almost certainly be subject to a strong economic influence, in particular the capital cost of the equipment will be balanced against the saving of time.

2.6 ACCURACY

The normal starting point for determining the accuracy required is a proportion of the specified tolerance. The simple rule is that measuring equipment should be capable of consistent resolution to an accuracy of about 10% of the tolerance. Anything better than this is likely to be extravagant in that the inspection cost, both in terms of the equipment and the time taken, will be higher than it need be. It may be possible, and indeed preferable, to use equipment which falls short of the 10% requirement provided that the consequences are recognised and taken into consideration.

A number of factors contribute to the accuracy of measuring equipment, particularly the resolution and the repeatability.

The resolution might be defined as the smallest difference between readings which can consistently be detected by the average inspector. Up to a point the resolution is enhanced by magnification. In fact, only in the simplest of instruments, like the basic rule, is magnification not found. It is generally achieved by mechanical or optical means, but it can be electrical or electronic, particularly for digital readout instruments. In the case of needle or column-type instruments, the magnification can be calculated by dividing the linear movement of the indicator against its scale by the movement of the measuring stylus needed to produce it. For example, if a measuring stylus movement of 0.1 mm appears as an indicator movement of 10 mm, then the magnification is 100. In the case of digital readout devices, the concept is not so simple, but an indication can be obtained from the order of accuracy to which the result is displayed. If the measurement requires movements of the order of 0.1 mm and the results are displayed to three decimal places, it can be assumed that the magnification is again about 100.

It might appear that resolution can be improved indefinitely merely by increasing magnification, but the magnification cannot usefully be increased beyond the point where readings do not consistently repeat. There is no value in magnification which gives readings to three decimal places if the basic design of the instrument is such that it can only be consistent to the second decimal place. Ideally an instrument might be expected to be the reverse, i.e. if the reading is to three decimal places, consistency of reading to the fourth decimal place is desirable. This, however, approaches perfection and in practice most instruments would be considered quite satisfactory if the reading consistency was half the last figure required, i.e. for a measurement of three decimal places, a consistency of reading to 0.0005 would be acceptable.

Another important factor influencing resolution is the ease of reading the instrument. This will depend, to quite a large extent, on its mechanical design and such details as the clarity of markings and colour of needles and scales. Accuracy will also depend on the precautions taken, both in the design of the instrument and by the user, to reduce or eliminate altogether problems which can give erroneous readings. This is particularly important at very high accuracies, and examples are distortion and parallax.

2.7 DISTORTION

Most of the inaccuracies arising from distortion will be avoided if Abbe's Law is continuously borne in mind. This states that maximum measuring accuracy is obtained if the axis of measurement is the same as the axis of the instrument. It is illustrated with a caliper type micrometer in Fig. 2.5.

14 Philosophy of Measurement

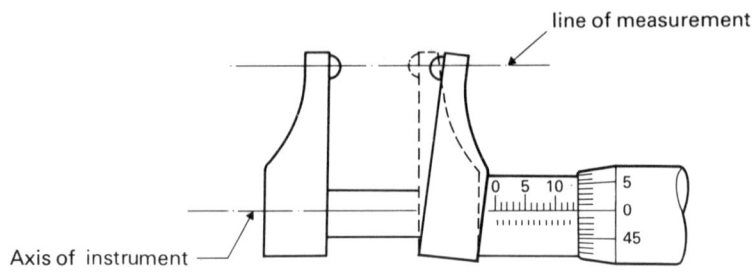

FIGURE 2.5

It will be seen that errors could arise both from distortion due to the measuring pressure applied, and from the possibility that the anvils might not move precisely parallel to each other. Figure 2.6 illustrated how some instruments, e.g. the normal micrometer, inherently satisfy Abbe's Law, whilst others, e.g. the caliper gauge, do not.

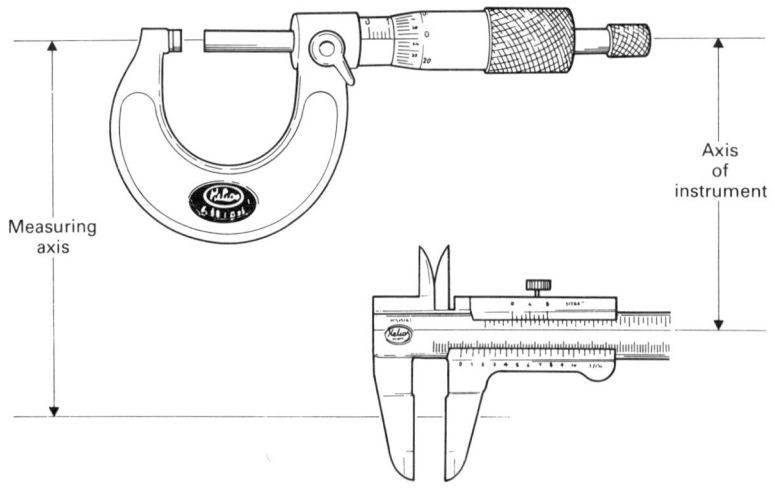

FIGURE 2.6

2.8 PARALLAX

This problem arises in the common case of scale graduations and datum lines against which they are read not being in the same plane. Figure 2.7 illustrates this situation on the sleeve and thimble of a micrometer, and it will be seen that a true reading will only be

Use of Comparison 15

obtained if the graduations are viewed in direction (b), which is along a radius of the micrometer barrel.

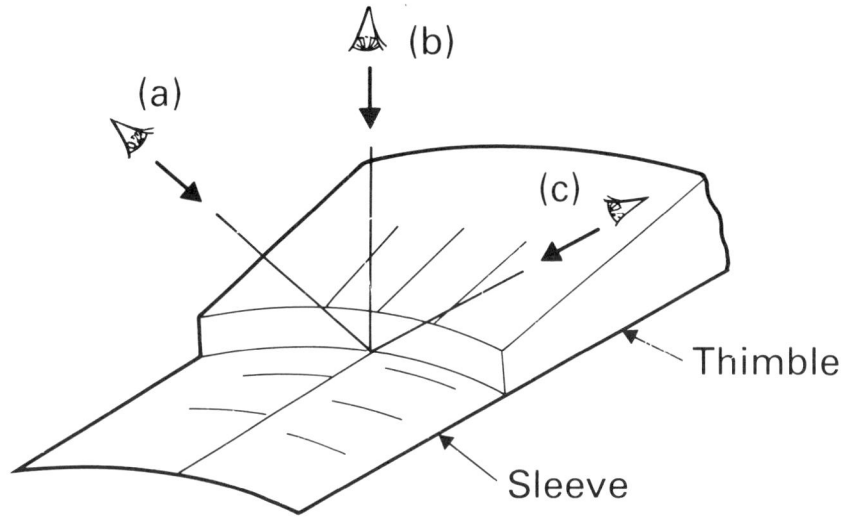

FIGURE 2.7

It is good practice to make a habit of always viewing scales in a direction which will reduce parallax to a minimum, even if the accuracy required in a specific application does not necessarily justify it.

2.9 GENERAL MEASURING ACCURACY

The pictorial diagram in Fig. 2.8 indicates the level of accuracy which might be expected from various instruments when carrying out commonly encountered measuring tasks.

2.10 USE OF COMPARISON

Other inaccuracies which might be due to the design or condition of instruments can be minimised or eliminated by using the principle of comparison, as described in Chapter 4, whenever possible. When this method is used, the most accurate results will be obtained if the dimensions being compared are as close to each other as is practicable.

16 Philosophy of Measurement

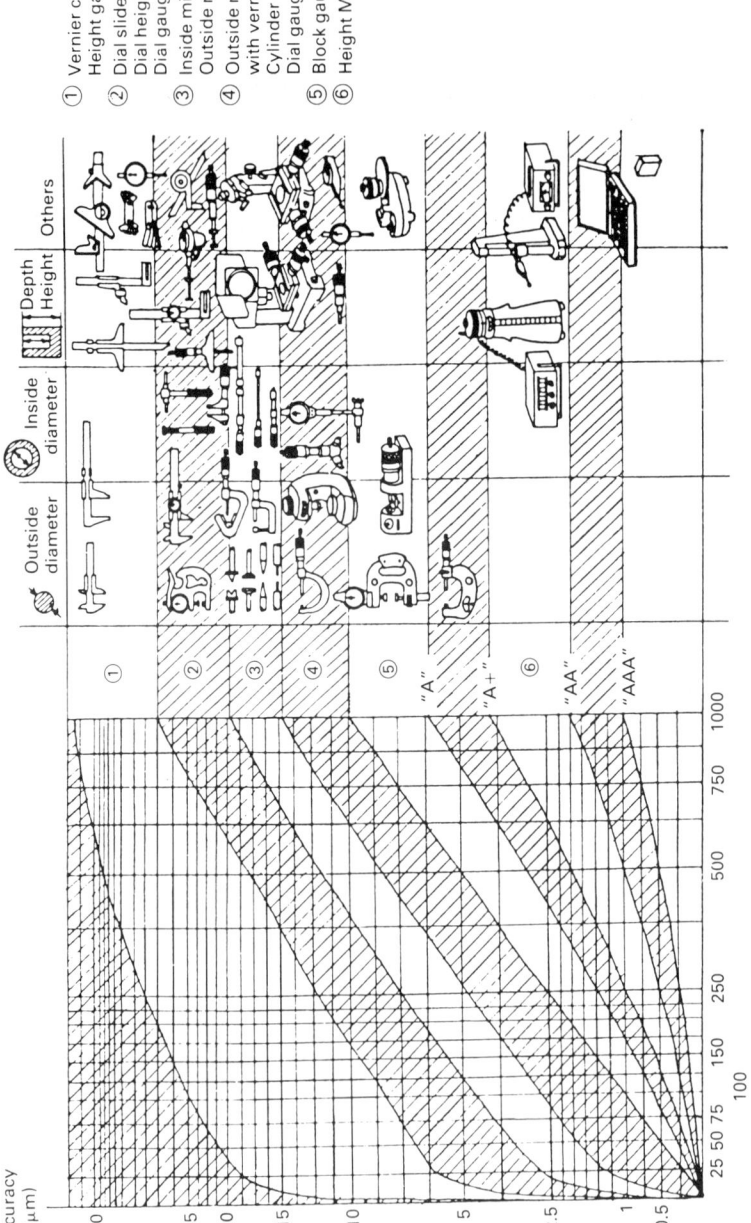

FIGURE 2.8

Chapter 3

Specifying the Requirement

In order to appreciate the objectives of an inspection activity fully, it is important that the philosophy of tolerances is understood.

A measurement of any sort will, as indicated in Chapter 2, almost certainly be made against a precise requirement. This is likely to be specified initially by the designer of the product, and to be precisely quantified, e.g. a shaft diameter of 20 mm.

It is, however, necessary to realise that nothing is ever absolutely precise, and, however meticulous the requirement, some margin must be allowed for deviation from the precise size.

This margin is known as the tolerance.

3.1 TOLERANCE PRINCIPLES

The tolerance can be specified in two basically different ways, either as a deviation in one direction only from the precise, or nominal, size, or it can be a deviation allowable in either direction from the nominal.

These two approaches are illustrated diagrammatically in Fig. 3.1.

Although guidelines can be given for selecting the system, they will not necessarily be appropriate in all circumstances.

Generally, a dimension to a surface not in contact with anything else can be given a tolerance on either side of the nominal and this is known as a bilateral tolerance. Dimensions to surfaces which have a fit which is significant to the performance of the product, e.g. shafts in holes, are likely to have a tolerance in one direction only, and this is known as a unilateral tolerance. Unilateral tolerances are usually arranged so that nominal size dimensions give the tightest fit.

This philosophy automatically gives a situation in which the nominal size is the "maximum material" condition, i.e. further material could be removed and still allow the dimension to be within the required tolerance. This concept is useful in that it encourages manufacture to be undertaken in a way which is least likely to result

18 Specifying the Requirement

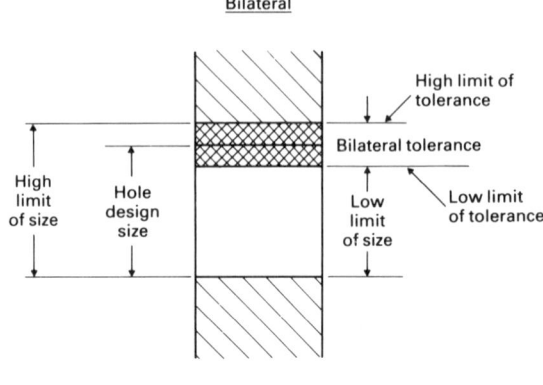

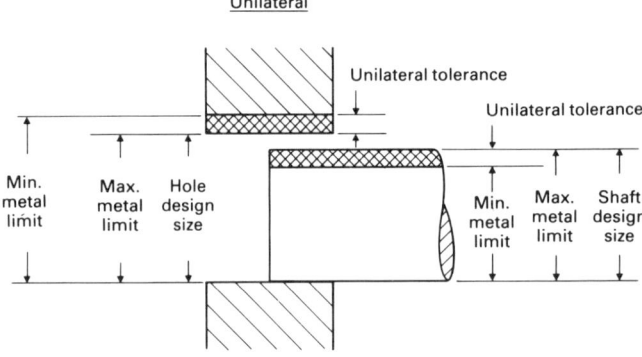

FIGURE 3.1

in scrap, because the tendency is for manufacture to aim at the nominal size. Normally, there will be some deviation on either side of the size aimed at, and, under these conditions, Fig. 3.2 shows how a nominal size on maximum metal condition is safer.

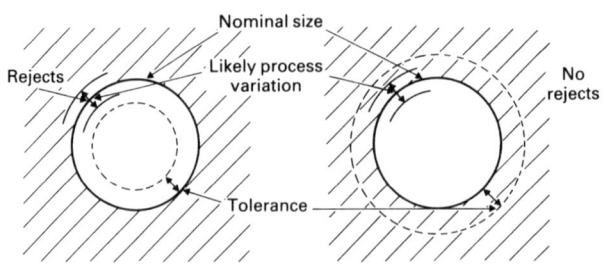

FIGURE 3.2

The reason can be summarised as the ease with which correction can be achieved by the removal of further material, compared with the considerable difficulty, even impossibility, of correction by adding material.

Having determined in principle whether a unilateral or a bilateral tolerance is required, it is then necessary to quantify the tolerance permissible. At this point there might be a conflict between the designer and the manufacturer. The designer will be tempted to "play safe" by quoting the smallest tolerance he thinks will be acceptable, whilst the manufacturer will be anxious to have as large a margin for error as possible.

The tolerance must, of course, be small enough to ensure acceptable and reliable performance of the product, but an excessively tight tolerance can result in considerable unnecessary expense. The selection of suitable manufacturing tolerances is of such significance that it could make the difference between the commercial success and failure of a product. It is desirable that encouragement to comment on and criticise tolerances is given, not only to design and manufacturing functions, but also to those involved in such activities as after sales servicing.

In order to ensure that advantage is taken of new developments in manufacturing techniques and of experience of the product in service, it is good practice to establish formal feedback systems which will indicate changes in tolerances likely to be of benefit to the commercial viability of the product.

In practice, it is rare for tolerances to be selected entirely from first principles. Two guidelines are generally available, the first is the precedent established by similar features of other products, and the second is the range of standard tolerance systems which exist.

As far as the first is concerned, most products are either developments of earlier models, or include features which have been used already in similar products, and it is obviously sensible to take advantage of any experience of these.

3.2 STANDARD TOLERANCE SYSTEMS

The second form of assistance can perhaps be best illustrated by examples.

A large proportion of tolerance decisions have to be made in connection with the fit of cylindrical features. These might generally be described as holes and shafts even if, as in the case of a bolt fitting into a hole, there is no relative movement between the parts when the product is in operation.

20 Specifying the Requirement

There are three ways in which a desired fit might be obtained. The first is that one of the parts, generally the hole, is made to any size within a fairly wide tolerance. This is then measured and the mating part is made to a size calculated from the actual size with a tight tolerance. This is an extravagant means of achieving the objective, and it results in the parts being subsequently not interchangeable. Whilst historically it was a common method, it is now only likely to be used in the case of large "one off" products.

The second method is to manufacture both mating parts within a tolerance which is too wide to give the desired fit automatically. This is then obtained by selecting from the range of parts produced, two which are of such a size that they give the fit required. This method is known as selective assembly. It is to be avoided because it involves either measurement of parts at the assembly stage or categorising the parts in terms of the size of a particular feature. It is sometimes, however, an economical way of obtaining a tight tolerance fit which would otherwise only be possible if the tolerances on the individual parts were very expensive to achieve.

The third way is the obvious one of giving each of the parts a tolerance which will result in an acceptable fit, regardless of where their individual sizes lie within the tolerance band. This situation is illustrated in Fig. 3.3 in which the desired fit lies between X and Y.

This method is much preferred to either of the other two and should be the objective wherever possible.

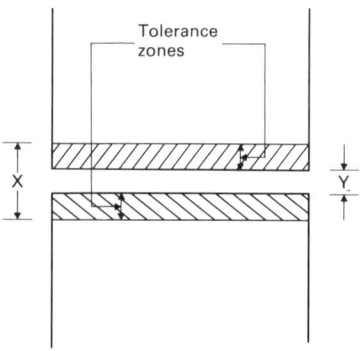

FIGURE 3.3

Having decided that this principle is to be adopted, the sizes and tolerances can be determined on either a "hole" or a "shaft" basis. These alternatives are illustrated diagrammatically in Fig. 3.4.

Standard Tolerance Systems 21

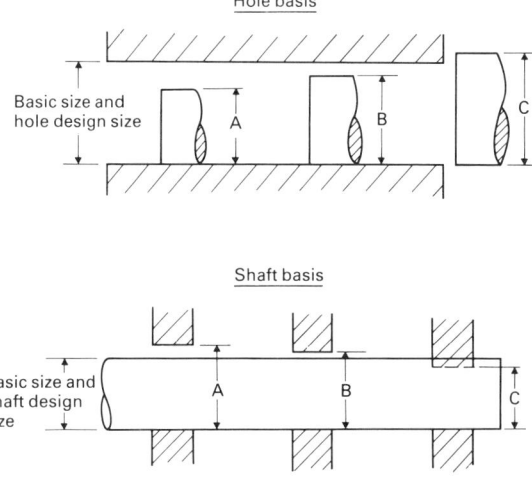

FIGURE 3.4

If the hole basis is used, the size of the hole is fixed as a datum and the shaft size is selected to give the required fit. For the shaft basis, the reverse is the case. It is normally more convenient for manufacture to determine the sizes using the hole basis.

Since this dimensioning of holes and shafts is so common, the task of selection has been simplified by nationally, and internationally, accepted tolerance systems. Such a system is illustrated in Fig. 3.5 and 3.6. This system is devised in such a way that, regardless of the size of the parts, a letter combination represents a type of operational fit.

Holes	Shafts					
	Clearance				Trsn.	Interf.
	c11	f7	g6	h6	n6	t6
H11						
H8				*		

FIGURE 3.5

22 Specifying the Requirement

The tabulation in Fig. 3.5 gives a selection of fits on the "hole basis", since seven tolerances are given for shafts but only two for holes. The last two columns are "transition" and "interference" fits. Transition gives a situation which could give either a clearance or interference, depending on the top position within their individual tolerances of the parts concerned. Both of these fits are used for location of parts relative to each other. Using this tabulation, a fit might be specified as, for example, H8-f7.

The dimensional value of the fit may be obtained from a tabulation such as Fig. 3.6.

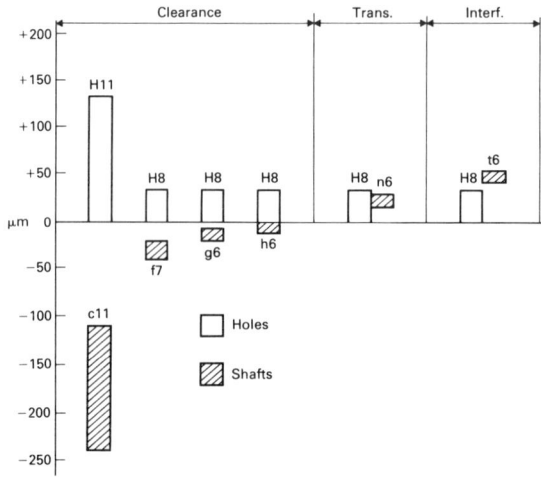

FIGURE 3.6

This gives the actual fit represented by the codes for holes and shafts with a nominal size of 25 mm.

This, however, still does not quantify the tolerance values and these have to be obtained from the tabulation in Fig. 3.7.

Whilst in the majority of cases tolerances can be determined using these principles, there will be certain circumstances, where conflicts exist between economical manufacture of parts and satisfactory function of the completed product.

A simplified example is shown in Fig. 3.8, in which a disc A, mounted on a shaft B, is located by bearing C in a diaphragm D. This, in turn, is bolted to two static casings E and F. It is possible that functioning of the mechanism will require a close tolerance on the clearance X between disc A and casing F.

Standard Tolerance Systems

Basic size		Holes		Shafts					
over	to	H11	H8	c11	f7	g6	h6	n6	t6
mm		Tolerance unit 0.001 mm							
		+	+	−	−	−	−	+	+
−	3	60 0	14 0	60 120	6 16	2 8	0 6	10 4	24 18
3	6	75 0	18 0	70 145	10 22	4 12	0 8	16 8	31 23
6	10	90 0	22 0	80 170	13 28	5 14	0 9	19 10	37 28
10	18	110 0	27 0	95 205	16 34	6 17	0 11	23 12	44 33
18	30	130 0	33 0	110 240	20 41	7 20	0 13	28 15	54 41
30	40	160 0	39 0	120 280	25 50	9 25	0 16	33 17	64 48
40	50			130 290					70 54

FIGURE 3.7

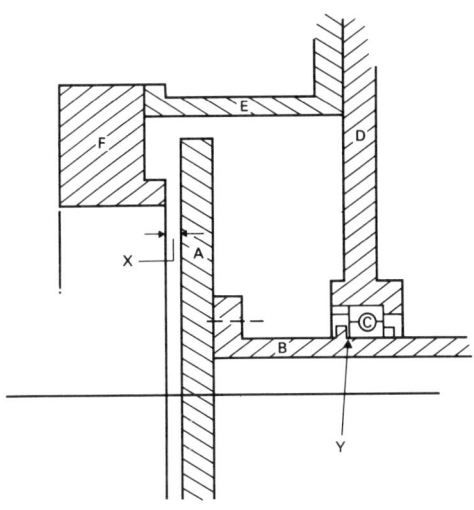

FIGURE 3.8

In this example, this clearance is determined by the tolerances on six dimensions and it is likely that, to achieve a close limit clearance consistently, excessively tight tolerances on the individual parts would be demanded. In such cases it may be expedient to introduce an adjusting washer in position Y, which will be machined to a size determined by a trial assembly, to give the required clearance X.

3.3 GEOMETRICAL TOLERANCES

In addition to straightforward linear dimensions, relationships between one surface and another which cannot be measured in this way are equally important. Examples of such relationships are parallelism and squareness. Traditionally, their accuracy has been dependent on the means of production, but it cannot be assumed that manufacturing methods will automatically give adequate results.

The obvious, and widely used, means of defining such requirements is to specify them in a narrative and qualitative way, e.g. "Surfaces marked thus to be parallel." Rather more precision can be obtained if something of a mixture of qualitative and quantitative methods is used. The narrative comment might then include a dimension, perhaps "Surfaces marked thus to be parallel within 0.1 mm."

If the number of characteristics to be defined is small, the second of these methods might be adequate, but there is scope for misinterpretation and, in the case of a complicated part, it could result in a drawing which is confused by wordy comments.

The increasing complexity of engineering specifications, and the desire for greater precision has provoked the development of systems which can quantify such geometrical characteristics precisely and, generally, symbolically.

Currently, the basis for most of these is an ISO (International Standards Organisation) system. There are still many systems, however, and care must be taken not to assume that a specific symbol will always mean the same thing.

As an example of the way in which such a system is used, Fig. 3.9 illustrates the definition of a roundness tolerance of 0.05 mm on a shaft.

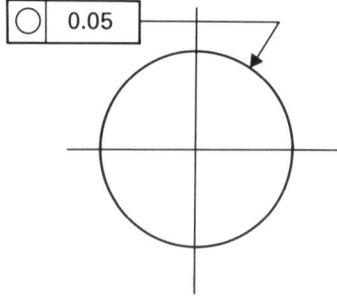

FIGURE 3.9

The symbol 0 is used to indicate roundness, and this is placed in a rectangular frame, together with the tolerance, and identified to the surface concerned by means of an arrow.

What this means in practice is that the surface of the shaft at any cross-section must lie within two concentric circles 0.05 mm apart, as in Fig. 3.10.

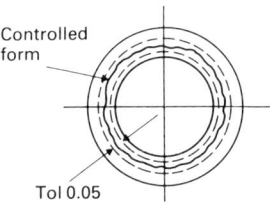

FIGURE 3.10

With this system, it is important to note that the difference between the diameters of the circles in Fig. 3.10 is 0.1 mm, and it is an indication of the confusion that still exists in the specification of geometric tolerances that those who have experience with other systems might interpret the symbols in Fig. 3.9 as a requirement for the shaft surface to lie within two circles different in diameter by 0.05 mm.

It is, of course, only necessary to quote a roundness tolerance if it is required to be less than the diameter tolerance, otherwise the diameter tolerance will automatically control the roundness to within the same figure. In fact, if no geometrical tolerance is quoted, it can be justifiably assumed that roundness within the diameter tolerance is acceptable, as depicted in Fig. 3.11.

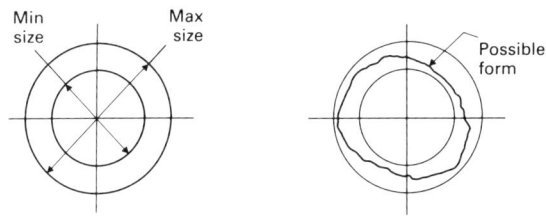

FIGURE 3.11

This philosophy also applies to other geometrical relationships. Parallelism, for example can be assumed to be acceptable within the tolerance on the dimension between the surfaces concerned, unless an additional tolerance requires otherwise.

Returning to circular parts, another common characteristic often requiring specific definition is concentricity. Defining this feature introduces the concept of nominating one feature as a datum. This is illustrated in Fig. 3.12 in which one of the axes is the datum and the other is given a positional tolerance with respect to it.

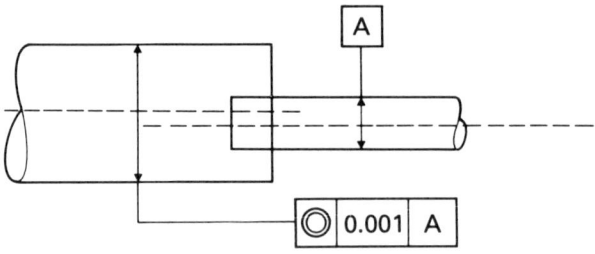

FIGURE 3.12

It is, of course, not only circular features that are involved in closely defined geometrical relationships. Figure 3.13 shows how the squareness of two surfaces nominally at right angles may be specified. This again involves nominating one of the surfaces as a datum and identifying it with a letter, in this case A.

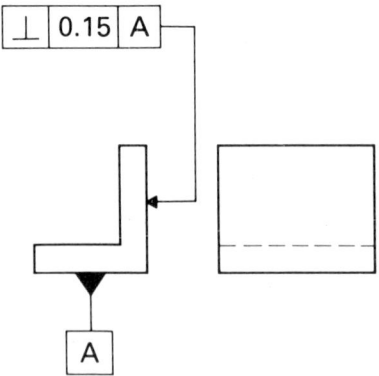

FIGURE 3.13

The squareness characteristic is denoted by the symbol ⊥ and Fig. 3.14 shows how this, together with the tolerance required, might appear on a drawing.

Process Capability 27

The interpretation of this form of presentation is that the whole of the vertical surface must be within two planes, precisely perpendicular to the horizontal surface, spaced 0.15 mm apart, as shown in Fig. 3.14.

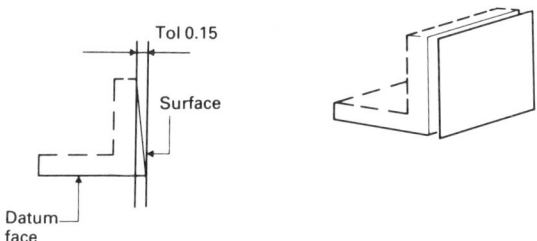

FIGURE 3.14

3.4 PROCESS CAPABILITY

When manufacture actually commences, formal assessment of the compatibility of the tolerance requirements and the manufacturing method can be obtained by measuring the process capability.

This is defined as the extent to which a manufacturing process consistently and economically satisfies the requirements of the specification. It can be practically determined by measuring the dimension concerned on a sample of the parts produced. Usually about twenty-five parts are sufficient to obtain a reasonable prediction of what will happen in long run production. Using recognised statistical techniques, a measure of the process capability can be obtained which might appear in graphical form as shown in Fig. 3. 15.

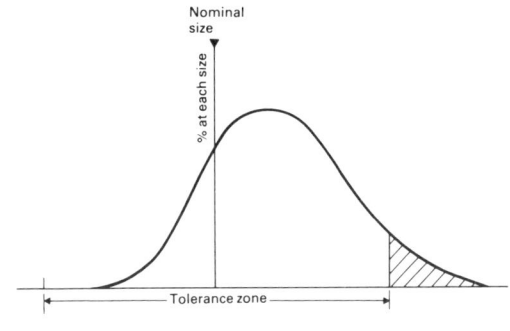

FIGURE 3.15

28 Specifying the Requirement

The vertical scale is the proportion of the total parts produced at each specific size. The whole area beneath the curve represents 100% and the proportion which will be outside the tolerance is represented by the shaded area.

The sample results can also be analysed to give a single figure for the proportion of parts which can be expected to be within the tolerance. This figure should be at least 97%, otherwise, bearing in mind the number of individual dimensions contained in a part which might all be subject to a similar degree of error, a large proportion of scrap will result.

Process capability is the ultimate test of the viability of specification and manufacturing methods. Unless it is adequate the economic success of a product must be suspect. It is, therefore, sound practice to carry out process capability studies automatically for all new processes and for significant changes.

Chapter 4

Basic Mechanical Measuring Instruments

The object of this chapter is to describe and discuss the function of basic, and primarily mechanical, instruments for linear measurement, and to give some indication of the large variety of measuring tasks which may be undertaken with comparatively simple equipment.

4.1 THE STEEL RULE

Perhaps the simplest measuring instrument used for engineering purposes is the steel rule. A reasonable quality rule can have marked divisions of 0.05 mm, and these can be read consistently to 0.02 mm, perhaps better than this if the rule is of high quality and a low magnification lens is used to read the scales.

Accuracy of this order can be expected if the rule is applied directly to the work, but the accuracy, and the versatility of the rule, can generally be increased if calipers are used to transfer the dimension from the work to the rule. Calipers are readily available in a wide variety of shapes and sizes, and they can be specially made to deal with dimensions which are not easily accessible with standard types. In their simplest form, they have friction hinges, but for engineering work the spring-loaded variety is much preferred. This type of caliper is set by the rotation of a knurled nut. A selection of steel rules and calipers is shown in Fig. 4.1.

An obvious development of this principle is to combine the steel rule and the caliper into one instrument. This is done by incorporating a fixed anvil into one end of the rule, and providing a second anvil which slides along the rule, as shown in Fig. 4.2.

This, by itself, improves the accuracy by providing a much more positive location of the scale against the work, but a further

30   Basic Mechanical Measuring Instruments

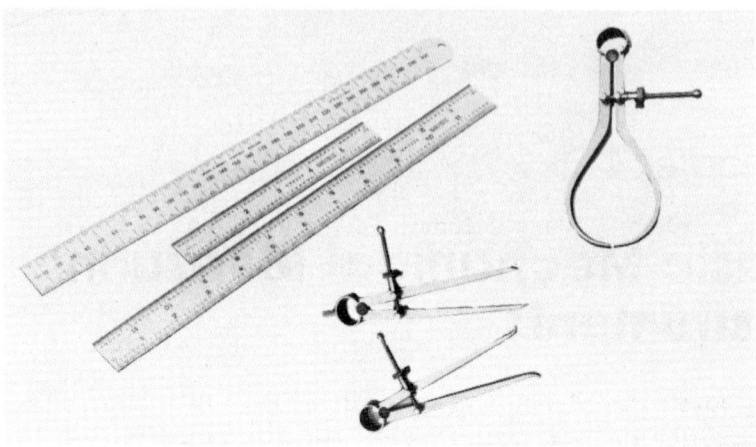

Figure 4.1

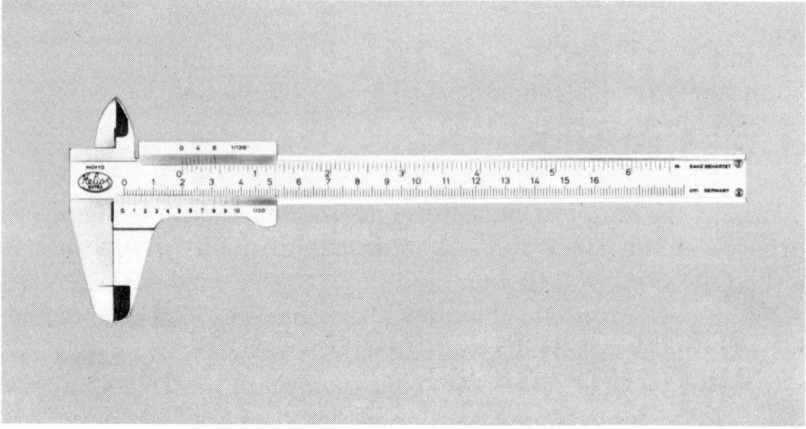

Figure 4.2

considerable improvement is obtained by including a vernier scale. Such a scale is incorporated in the instrument shown in Fig. 4.2 which is, in fact, known as a vernier caliper.

4.2 VERNIER SCALES

The vernier scale is a second scale attached to the moving anvil and in sliding contact with the main scale. This second scale is divided into units which are slightly smaller than those in the main scale. In the example shown in Fig. 4.3 the vernier scale is 49 mm

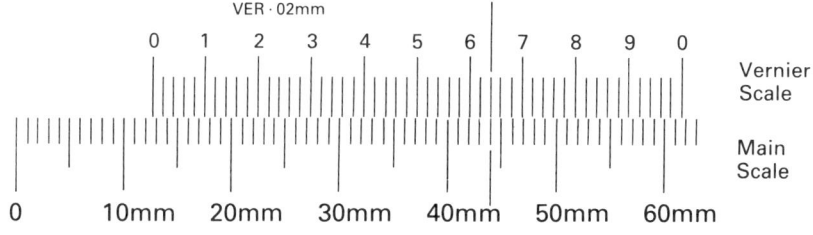

FIGURE 4.3

long but it is divided into 50 units. Thus each unit is 49/50 of, or 0.02 mm shorter than, the main scale. Because the difference is 0.02 mm, the vernier divisions are grouped in fives, each group representing a difference of 0.1 mm.

In the example, the position at which the zero on the vernier scale intersects the main scale is the dimension required. It will be seen that this is a little over half-way between 12 and 13 mm. The distance from the 12 mm mark can be accurately read, in this case to 0.02 mm, by looking along the vernier scale until a point is found at which the markings on both scales coincide. In Fig. 4.3 it is at 6.4. This represents 6.4 x 0.1 = 0.64 mm, and thus the measurement required is 12 + 0.64 mm or 12.64 mm.

4.3 VERNIER CALIPER, HEIGHT GAUGE AND DEPTH GAUGE

The combination of integral fixed anvil and scale with moving anvil and vernier scale is the basis of several very useful instruments.

The vernier caliper has already been mentioned, and this is available in a large range of sizes.

If the instrument is mounted at right angles to an accurately machined base, which then takes the place of the fixed anvil, it will then measure the distance from the base on which it is mounted to the face of the moving anvil. The instrument has now become a height gauge, as the example shown in Fig. 4.4.

The measuring scale of both these instruments is, of course, of very much more robust construction than the steel rule, and the squareness, parallelism and finish of measuring and locating faces are such that measurements can consistently be made to an accuracy of around 0.15 mm per metre.

32 Basic Mechanical Measuring Instruments

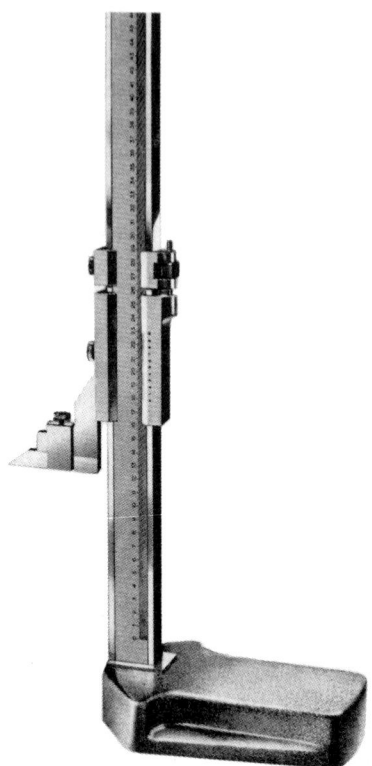

Figure 4.4

It will be noted that this particular instrument has an additional feature in that the anvil and moving scale are linked to a clamp, which also slides along the main scale, through a threaded rod and knurled nut. This is of particular value in setting the gauge to a predetermined dimension. In use the anvil and clamp are moved together until the anvil is in approximately the correct position. The clamp can then be locked to the main scale and the movement of the anvil finely controlled by rotation of the nut.

The vernier caliper and vernier height gauge are very widely used and can, with ingenuity, accomplish a very large proportion of measuring tasks.

The depth gauge shown in Fig. 4.5 is another popular variation of a linear measuring scale with a vernier reading and specialised anvils. This is also an instrument with many applications and is capable of measuring to an accuracy of 0.05 mm.

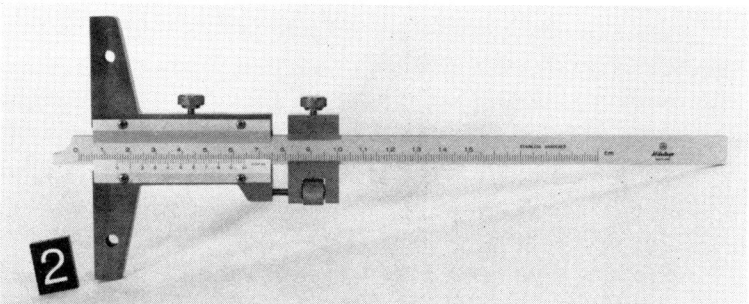

FIGURE 4.5

The height gauge is perhaps the most versatile of these three instruments, but it can only be used with a flat surface on which the height gauge and the part to be measured can be located.

4.4 SURFACE TABLES, BOX CUBES AND VEE BLOCKS

Traditionally the flat surface table will be either a free standing cast iron surface table or a smaller cast iron plate suitable for mounting on a bench. Both will have their upper surface hand scraped to a very high standard of flatness. The level of flatness of a particular table may be obtained from its grading for which there is an internationally recognised standard. An indication of the actual flatness values may be obtained from the tabulation given in Fig. 4.6.

The interpretation of the values is that, for example, all points on a grade 2 table 8' x 4' will be contained between two parallel planes 0.0015" apart. For general production and inspection use surface tables of grades 2 and 3 are quite adequate.

Imperial Sizes	Grade 1	Grade 2	Grade 3	Marking Out
3' x 2'	0.00045	0.0009	0.0018	0.004
3' x 3'	0.00050	0.0010	0.0020	—
4' x 3'	0.00050	0.0010	0.0020	0.004
5' x 3'	0.00055	0.0011	0.0022	0.006
6' x 3'	0.00060	0.0012	0.0024	0.006
6' x 4'	0.00065	0.0013	0.0026	0.006
8' x 4'	0.00075	0.0015	0.0030	0.007
8' x 6'	0.00080	0.0016	0.0032	—
10' x 6'	0.00090	0.0018	0.0036	—
12' x 6'	0.0011	0.0022	0.0044	—

FIGURE 4.6

Examples of a cast iron surface table and surface plates are shown in Fig. 4.7.

FIGURE 4.7

In recent years natural rock materials have become increasingly popular and granite tables are rapidly becoming the standard.

There are several advantages of granite over cast iron. In particular, it is remarkably stable, does not suffer from corrosion and, if damaged, burrs which can give misleading results are not formed. Figure 4.8 shows examples of a granite surface table and surface plate.

Since the height gauge will only measure in a direction perpendicular to the surface on which it is placed, the part must be located

FIGURE 4.8

on the surface table in such a way that the dimension to be measured is in this direction.

Having obtained the height gauge and the flat surface, the next requirement is the means of locating the part to be measured. Simple and very useful accessories for this purpose are the box cubes and angle plates shown in Fig. 4.9.

FIGURE 4.9

These are accurately machined castings, again obtainable in recognised accuracy grades, with provision for mounting work pieces using simple clamps.

Whilst box cubes and angle plates will cope with the mounting of most static parts, which generally have at least one flat mounting surface, they are not normally suitable for locating rotating parts, which are likely to be basically cylindrical in section.

For this purpose, the simplest, and perhaps most useful, accessory is a pair of vee blocks, as shown in Fig. 4.10.

FIGURE 4.10

These are invariably sold in matched and identified pairs, and it is essential that they are used in pairs. Fig. 4.11 shows a simple set-up of surface table, vee blocks and height gauge.

FIGURE 4.11

Vee blocks are also obtainable in a wide variety of sizes and styles and Fig. 4.12 shows a granite vee block together with a variation which makes use of balls to achieve the same results. The ball type has the advantage of easy rotation of the part being measured.

FIGURE 4.12

4.5 DISPLAY SYSTEMS

The reading of an index mark against a linear scale is the obvious and traditional method of obtaining a linear dimension. Most types of instrument are, however, available with the alternative analogue display of a circular dial, and also with mechanical or electronic digital display.

A most useful, and reasonably priced, instrument combines the dial readout with the caliper. The dial caliper has the same versatility as the simple caliper, in that it can be made in almost infinite variety to cater for otherwise inaccessible dimensions, and has the advantage of a direct reading without the need to compare against a scale.

A selection of dial calipers is shown in Fig. 4.13 and it will be apparent that the accuracy will depend on the length of the legs. A dial caliper should be selected with legs as short as possible and they should be used as comparators by setting against a standard, such as a cylinder or internal ring, which is close to the size of the dimension to be measured.

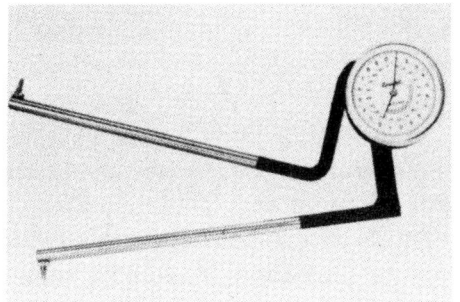

FIGURE 4.13

Used in this way, dial calipers are capable of consistent measurement to 0.05 mm.

Fig. 4.14 shows a further development of this instrument to incorporate an electronic digital readout. In addition to ease of reading, the same instrument can display either metric or imperial units.

Height gauges and vernier calipers are also available with the same alternative display systems, and Fig. 4.15 illustrates each of these instruments with the dial display.

38 Basic Mechanical Measuring Instruments

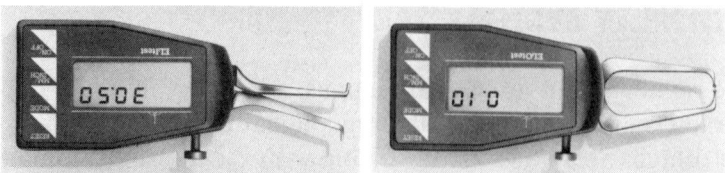

Figure 4.14

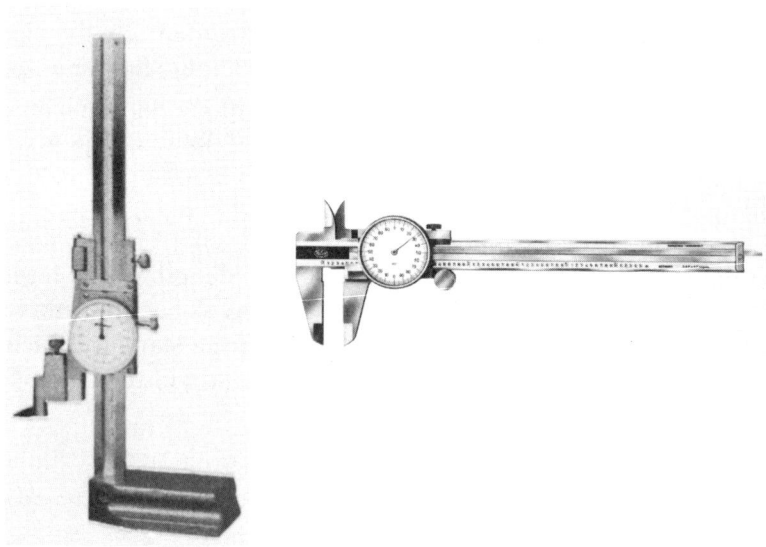

Figure 4.15

The digital readout alternative is shown, again for both types of instrument, in Fig. 4.16.

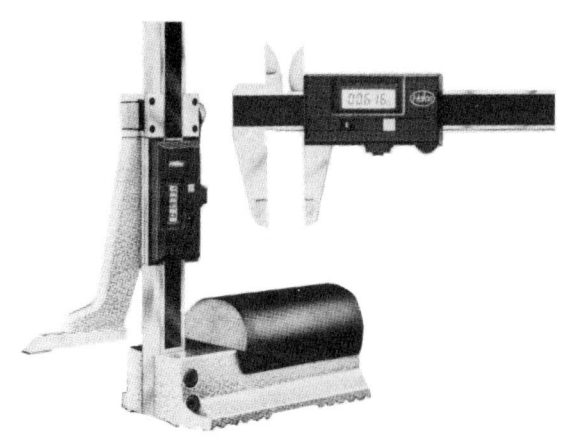

Figure 4.16

The digital instruments can give either metric or imperial dimensions, they can be set at zero in any position and they can read positive in either direction. Another valuable feature is that, in many cases, the readings can be obtained as an electrical output and processed by computer. The use of this facility is discussed in more detail in Chapter 12.

4.6 THE DIAL INDICATOR

So far, in using the height gauge and surface table, it has been assumed that the measurement has been obtained by contact between the workpiece and the anvil on the height gauge, as in the example of Fig. 4.11. In practice this is not a very satisfactory method, since it is difficult to determine the precise point at which contact is made. It is also not easy to be satisfied that the base of the height gauge is completely located on the surface table, and that the workpiece is not disturbed in its location.

A much better method is to use some device for identifying the position of the surface on the workpiece, and then transferring it to the anvil of the height gauge. An instrument commonly used for this purpose is the dial indicator. These are available in many styles and two examples are illustrated in Fig. 4.17.

These examples are both mechanical and have quite simple movements as shown in Fig. 4.18.

A rack and pinion converts the linear movement of the stylus into rotation and further gearing gives the required magnification.

A dial indicator with a different configuration, known as the lever type, is shown in Fig. 4.19.

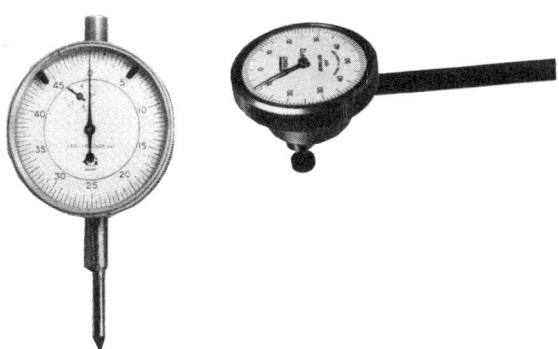

Figure 4.17

40   Basic Mechanical Measuring Instruments

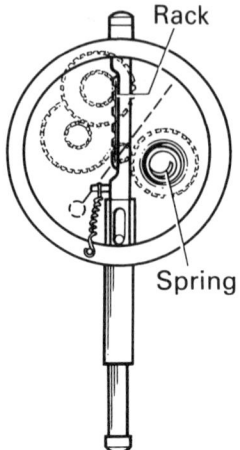

FIGURE 4.18

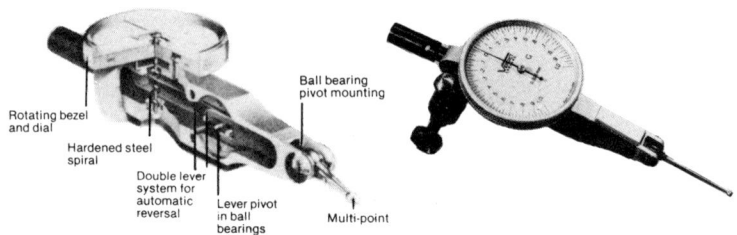

FIGURE 4.19

This form of indicator is particularly convenient for use with a height gauge, but care must be taken to see that the surface being measured is parallel to the measuring lever, since only in this position does the indicator accurately record the movement of the measuring head. Fig. 4.20 shows that, at any other position, the stylus length is effectively reduced.

FIGURE 4.20

An ingeniously shaped stylus, shown in Fig. 4.21, is available to minimise this effect.

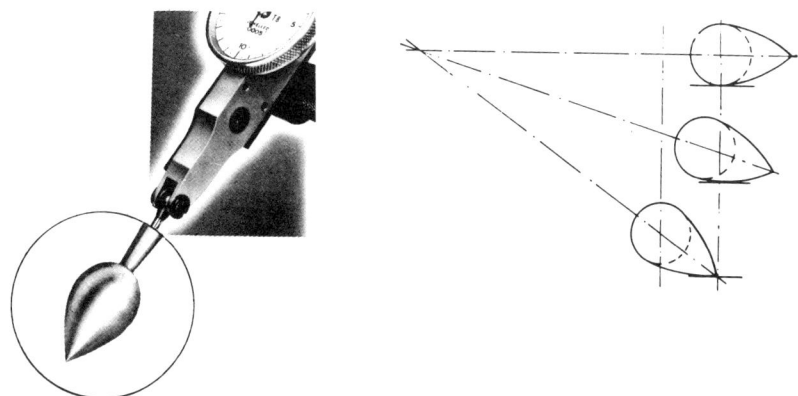

FIGURE 4.21

The dial indicator may be used either for quantitative measurement or for zero setting. In either case it is used to compare the workpiece surface with the measuring instrument surface.

This principle of comparison is most important and is illustrated diagrammatically in Fig. 4.22.

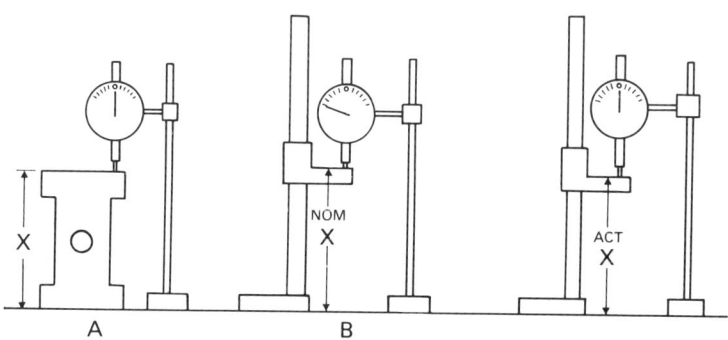

FIGURE 4.22

In diagram A the workpiece is mounted on the surface table in such an attitude that the dimension required is the vertical distance between the table and a surface on the workpiece parallel to it. The dial indicator is mounted on a stand with the measuring stylus in contact with the workpiece surface. It is desirable that the dial

42 Basic Mechanical Measuring Instruments

indicator is positioned so that the measuring stylus contacts the workpiece somewhere near the middle of its movement range, and preferably with the needle approximately vertical. In this position, the bezel of the indicator is turned so that the needle is at zero.

In diagram B the height gauge is set so that the upper surface of its anvil is at the nominal size of the dimension to be measured. The dial gauge and stand are then transferred to the height gauge with the measuring stylus in contact with the upper surface of the anvil. The reading on the dial indicator will then be the deviation on the workpiece dimension.

An alternative approach is depicted in diagram C, in which, having placed the measuring stylus in contact with the height gauge anvil, the height gauge is adjusted until the dial indicator again reads zero. The actual value of the workpiece dimension can then be read from the height gauge.

In order to use a dial indicator in this way, it must be mounted on a stand. It is not uncommon to make use of the instrument shown in Fig. 4.23, known as a surface gauge or scribing block, but the stability of this device is such that it should only be used if accuracy is comparatively unimportant, say ± 0.05 mm.

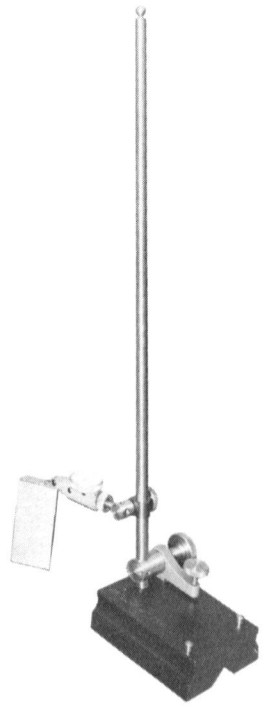

FIGURE 4.23

It is also not uncommon to mount the indicator on a height gauge of the type shown in Fig. 4.4, but this is extravagant because it makes little or no use of the height gauge's own measuring system.

The ideal indicator stand is made specially for the purpose, and an example is shown in Fig. 4.24.

Figure 4.24 also shows indicator stands with magnetic and pneumatic bases which enable them to be rigidly attached to a surface.

FIGURE 4.24

4.7 ELECTRONIC AND PNEUMATIC INDICATORS

The dial indicator also has its electronic counterpart and examples are shown in Figs. 4.25 and 4.26.

Figure 4.25 illustrates a digital version, whilst Fig. 4.26 combines digital and analogue displays.

These indicators have all the features already mentioned in connection with electronic displays, and can also have the considerable added advantage of a reading head which is remote from the measuring stylus, as shown in Fig. 4.27.

Another method of quantifying the movement of a measuring stylus is the pneumatic gauge, which differs in principle from both mechanical and electronic gauges, but again has the advantage of a flexible link between measuring stylus and reading head.

Historically, the pneumatic gauge appeared as an improvement on the mechanical indicator, particularly for multi-dimension applications. Its main disadvantage is that it requires a precisely controlled air supply, and has a very small measuring range. It has

44 Basic Mechanical Measuring Instruments

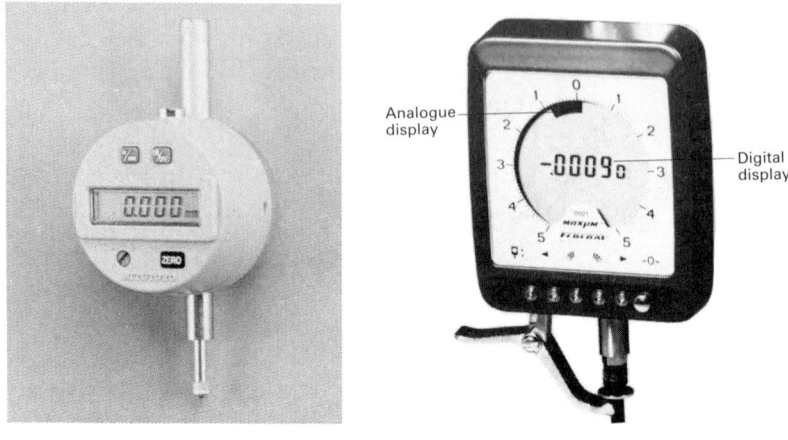

FIGURE 4.25   FIGURE 4.26

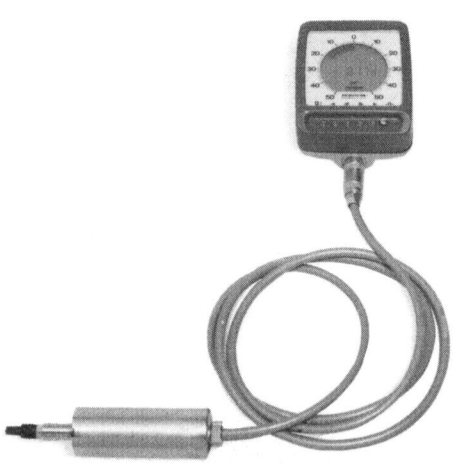

FIGURE 4.27

largely been superseded by the electronic indicator for measurements on flat surfaces, but it is still valuable as a means of measuring holes, particularly because, in this application, it is a non-contact method.

An example of a pneumatic gauge head for checking a hole diameter is shown in Fig. 4.28, together with two diagrammatic representations of operating systems.

The simplest system, Fig. 4.28(a), measures the gap between the head and the bore by the air pressure loss due to leakage through it.

Electronic and Pneumatic Indicators 45

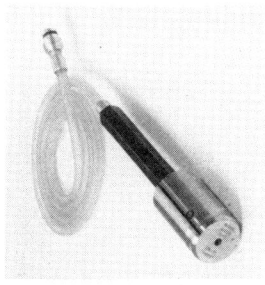
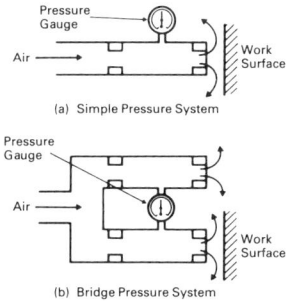

FIGURE 4.28

Another arrangement, the bridge pressure system shown in Fig. 4.28(b), measures the pressure difference between the loss through the gap and the loss from a completely open jet.

The non-contact characteristic of this method can be valuable if the bore has a high finish which might be damaged by contact methods described later in this chapter.

The most attractive method of obtaining an analogue reading from a pneumatic system is for the pressure to be measured by a simple liquid manometer. This form of presentation is shown in Fig. 4.29 and it will be seen that it lends itself to stacking for multi-dimension devices mentioned in Chapters 9 and 14.

The vertical column display is, in fact, so appealing that electronic simulations are in widespread use on non-pneumatic gauges.

FIGURE 4.29

4.8 THE MICROMETER

For many years the basis of engineering measurement was the ubiquitous micrometer, and it is only recently that its usefulness has been challenged by electronic and optical devices.

The micrometer, as shown in Fig. 4.30, is essentially an instrument which measures between two anvils, one of which is fixed and the other moves by rotation along an accurately produced screw thread.

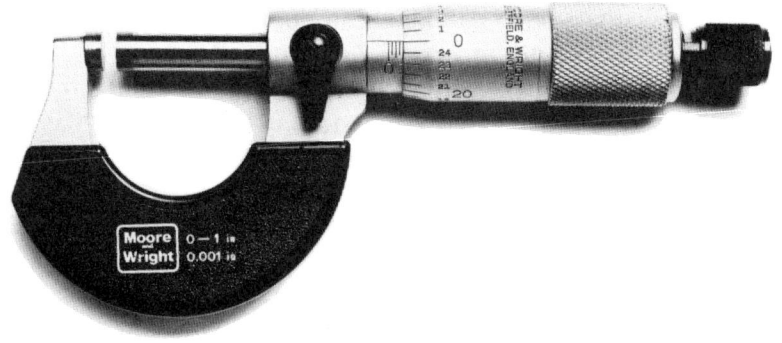

FIGURE 4.30

The moving anvil is rigidly attached to a thimble graduated so that its angle of rotation can be measured against a datum on the fixed body of the micrometer. If the thimble is rotated one complete revolution, then the moving anvil will move relative to the fixed anvil by an amount exactly equal to the pitch of the thread. The secret of the micrometer's precision is that the pitch can be accurately divided into small parts by measuring the angle of rotation on the thimble. In the case of a metric micrometer, the pitch of the thread is likely to be 0.5 mm and the circumference of the thimble divided into 50 parts. Thus each division on the thimble will represent a length difference of 0.01 mm. The reading might appear as in Fig. 4.31, and the actual dimension between the anvils is obtained by adding together the readings on the axial fixed scale and the circumferential moving scale.

The reading in this example is 7 + 0.37 = 7.37 mm.

Since the axial datum line is between the 37 and 38 graduations on the thimble it is, of course, possible to estimate the third decimal place with a consistent accuracy of perhaps 0.001 mm. The

The Micrometer 47

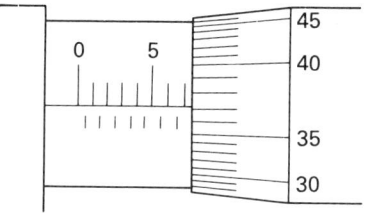

FIGURE 4.31

estimated reading to three decimal places in the example might be 7.373 mm.

The last decimal place estimate can, however, be made more reliably if the micrometer incorporates a vernier scale. In principle, such a scale is exactly the same as for the vernier caliper, but it is scribed circumferentially around the static part of the micrometer head, as shown in Fig. 4.32.

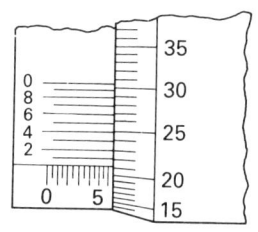

Sleeve: 6mm
Thimble: 0.21mm
Vernier: 0.003mm
―――――――――――――
Total: 6.213mm

FIGURE 4.32

The reading in this case is $6 + 0.21 + 0.003 = 6.213$ mm.

Figure 4.33 shows a cross-section of a typical micrometer barrel. Two particular features are a lock to hold the micrometer at a specific setting, at A, and a device for removing the effect of any clearance between the threads at B.

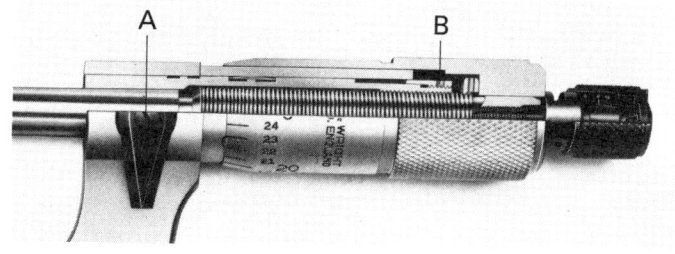

FIGURE 4.33

Micrometers, all using the same principle and effectively having the same measuring head, are available with an almost endless variety of frames. Simple frames for external measurement may be obtained with capacities up to 1 m, as the examples in Fig. 4.34 show.

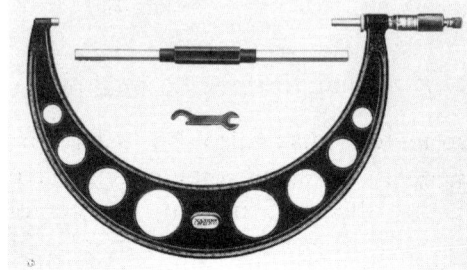

FIGURE 4.34

The range of the measuring head itself rarely exceeds 25 mm or 1 in, and if the micrometer is required to measure large dimensions, it must be used as a comparator and set to either a zero position or, preferably, to a dimension near the workpiece size, using some form of master. Larger micrometers are normally supplied with their own end bars, but for smaller sizes gauge blocks may be used.

Micrometer sets, as shown in Fig. 4.35, may be obtained to cover a wide size range, and, for economy, adjustable micrometers with interchangeable fixed anvils, shown in Fig. 4.36, can be used.

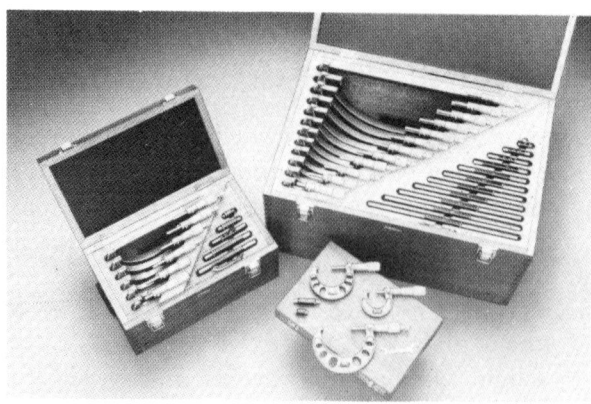

FIGURE 4.35

The Micrometer   49

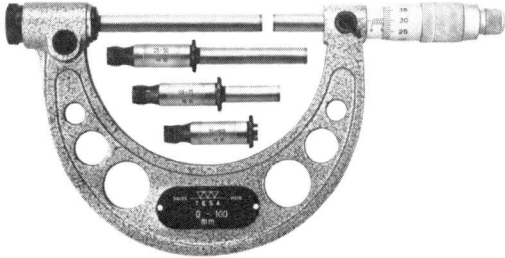

FIGURE 4.36

Micrometers may also be used for internal measurements. Instruments for small dimensions have jaws, whilst those for dimensions above 25 mm are in the form of a "stick" with the measuring head at one end. Examples of both types are shown in Fig. 4.37.

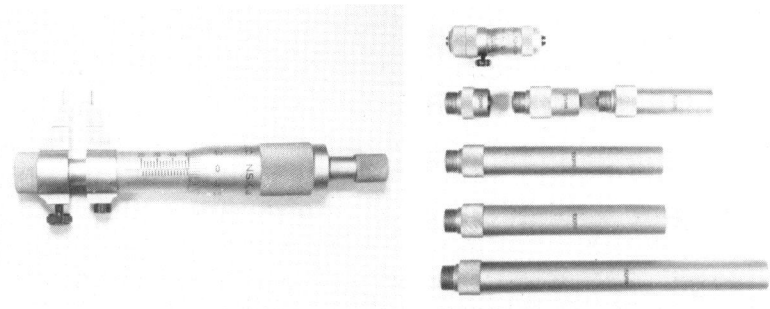

FIGURE 4.37

As in this example, "stick"-type internal micrometers are invariably supplied with sets of extensions which permit one head to cover a wide measuring range.

For greater ease of use, particularly when required for repetitive work, micrometers may be bench mounted. These also frequently incorporate a larger thimble, which permits greater accuracy, as in Fig. 4.38. A useful alternative for less frequent use of a micrometer in this manner, is the bench clamp shown in Fig. 4.39.

The bench micrometer shown in Fig. 4.38 also illustrates the use of a dial indicator actuated by the "fixed" anvil. Most micrometers incorporate a ratchet device to prevent an excessive load, with consequent distortion, being applied during measurement, and the dial indicator is a refinement which ensures that the measuring

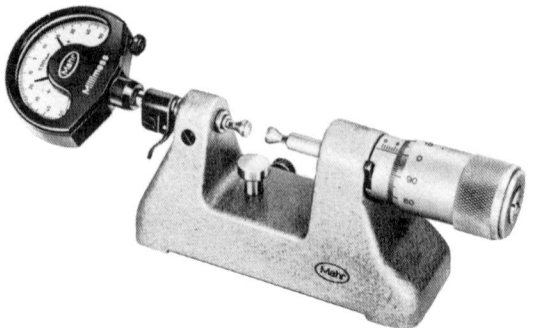

FIGURE 4.38

FIGURE 4.39

pressure is at a uniform and predetermined level. In use, the thimble would be turned until the indicator reading is zero, and the measurement then taken from the thimble position.

Many micrometers are manufactured with special frames for specific tasks. A few examples are shown in Fig. 4.40, but if a difficult measuring problem is encountered, a browse through the micrometer manufacturers catalogue might well reveal a solution.

Inevitably, micrometers are also available in digital form, and a mechanical example is shown in Fig. 4.41. The mechanism of such instruments is somewhat complex, and consequently expensive, and the electronic version shown in Fig. 4.42 is probably the more promising and economical development.

Electronic instruments again have typical electronic features, and comprehensive systems, based on digital electronic micrometers, as Fig. 4.43, may be obtained. These systems will measure, permanently record, and statistically analyse the results.

The Micrometer 51

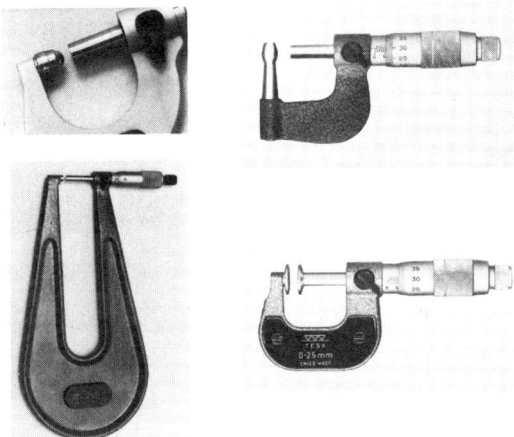

FIGURE 4.40

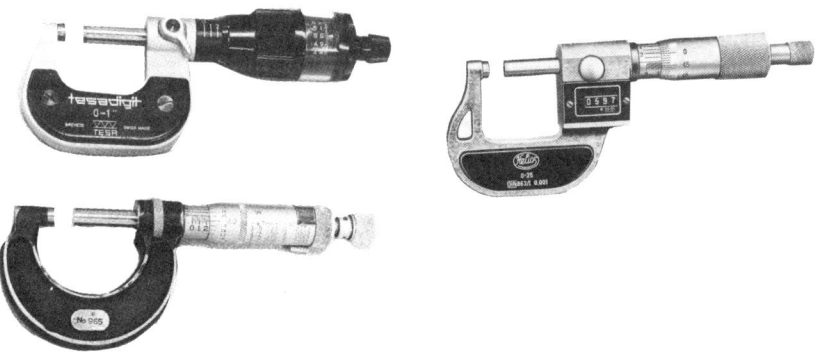

FIGURE 4.41

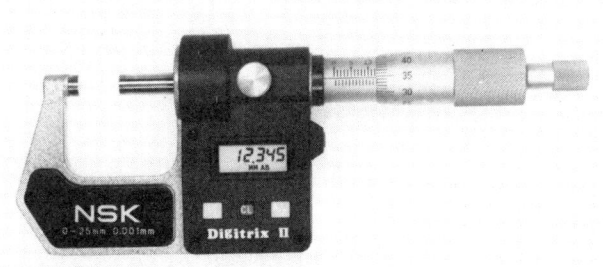

FIGURE 4.42

52 Basic Mechanical Measuring Instruments

FIGURE 4.43

A popular relative of the micrometer, which is used in the same context as a height gauge, is shown in Fig. 4.44.

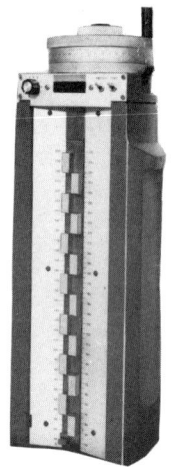 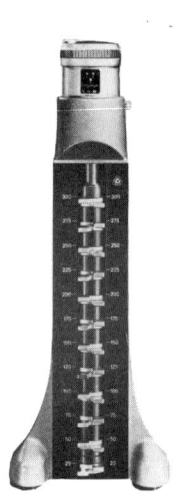

FIGURE 4.44

It is generally known as a height micrometer and consists of a number of very accurately machined horizontal datum faces, normally spaced 25 mm or 1 in apart, which can be moved vertically by means of a large diameter micrometer thimble at the top of the instrument. This enables the datum faces to be readily set to 0.002 mm.

4.9 BORE GAUGES, PLUG GAUGES

Reference has already been made to the measurement of holes using an internal micrometer.

The measurement of holes, particularly of small diameter, is such a frequent requirement that a large variety of suitable equipment is available.

Bore Gauges, Plug Gauges 53

A major disadvantage of the simple micrometer is that it measures across a diameter. Without taking a number of readings on the same bore, it is difficult to detect errors in the shape of the bore, of which ovality and lobing, as shown in Fig. 4.45, commonly appear, separately or in combination.

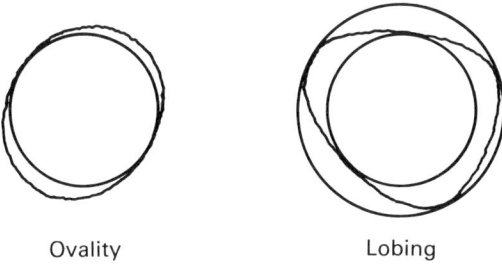

Ovality   Lobing

FIGURE 4.45

A development of the micrometer which overcomes this problem to a considerable extent is the bore micrometer shown in Fig. 4.46.

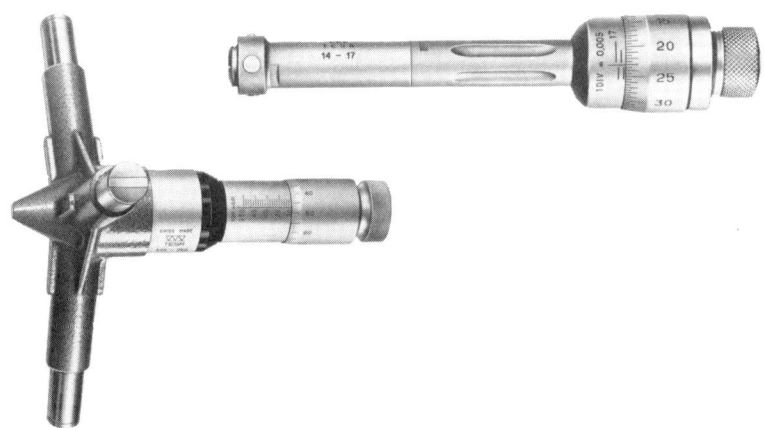

FIGURE 4.46

These instruments give a reading which is the mean diameter from three contact points. They come in a wide variety of sizes and can easily be read to 0.002 mm.

If a quantified measurement is required but this level of accuracy is not necessary, bore gauges of the type shown in Fig. 4.47 may be used.

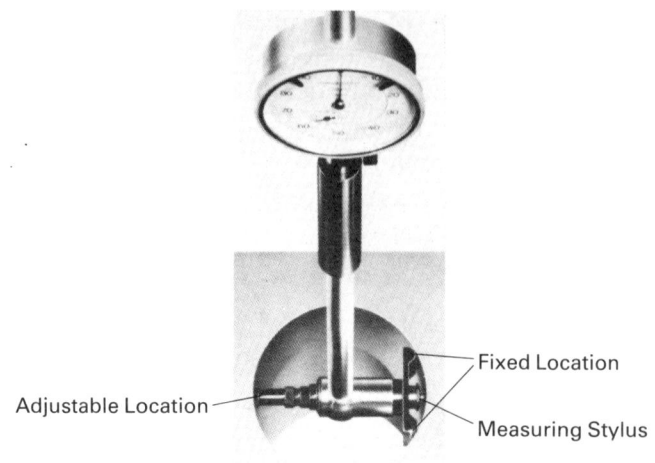

FIGURE 4.47

Such gauges mechanically transmit a diameter reading to a dial indicator. Although, as in the case of the internal micrometer, the measurement is only of a simple diameter, the three point location of the gauge ensures that it is positioned on a diameter in a radial plane to an accuracy as high as the bore itself. This means that the gauge has only to be "rocked" in an axial plane to obtain the correct reading. The "stick" micrometer, on the other hand, requires considerable skill to position it in both planes to ensure that a true reading is obtained.

Another advantage of this type of gauge is that the reading is displayed well outside the bore being measured. This greatly facilitates the measurement of deep bores.

If a quantified measurement is not required and it is sufficient to determine whether or not a hole is within its specified tolerance, by far the simplest and most economical device is the plug gauge. Examples are shown in Fig. 4.48, and they are normally used as a pair consisting of a "Go" and a "No-go" plug.

As these words imply, the tolerance of the gauges are such that if it is possible to insert the "Go" end into the hole, but it is not possible to insert the "No-go" end, then the hole is likely to be within the specified tolerance. The proviso is that the "No-go" end will react to the smallest diameter and it will not detect ovality. On the other hand, since most holes will, in the finished assembly, be fitted with either static bolts or studs, or with moving shafts, these gauges approach a functional check, which is one of the ideals of any inspection method.

Ring Gauges, Snap Gauges 55

FIGURE 4.48

4.10 RING GAUGES, SNAP GAUGES

The equivalent of the plug gauge for external diameters is the ring gauge shown in Fig. 4.49. Again, they are usually supplied in pairs consisting of a "Go" and a "No-go" and, in principle, they are used in exactly the same way as plug gauges.

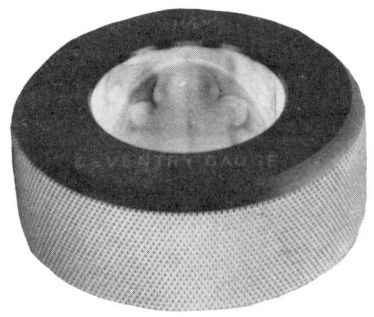

FIGURE 4.49

They are, however, not as valuable and not as widely used as plug gauges for several reasons. Particularly, they require more skill to handle, they can only be used in easily accessible positions and, in any case, external diameters are fairly easy and economical to measure by quantitative methods.

A popular "Go/No-go" alternative for volume production is the so-called snap gauge, illustrated in Fig. 4.50.

These consist of a frame in which is mounted a pair of anvils, the distance between the outer pair being the "Go" size, and between the inner pair, the "No-go" size.

56 Basic Mechanical Measuring Instruments

FIGURE 4.50

Snap gauges are made in both rigid and adjustable forms, both of which are illustrated in Fig. 4.50. The rigid variety is obviously cheaper to purchase, but the adjustable type have the advantages of being able to accomodate a range of dimensions and wear.

A development of the snap gauge is shown in Fig. 4.51.

FIGURE 4.51

This effectively replaces one of the anvils by a dial indicator and makes the gauge a quantitative instrument. The anvils on this type of gauge are normally fairly large in area, as in this illustration, and they are fitted with an adjustable stop at right angles to the anvils. This facilitates rapid inspection of diameters.

A rather more sophisticated type, with "built-in" dial gauge, is shown in Fig. 4.52, whilst Fig. 4.53 illustrates the ultimate, an electronic, digital read-out version.

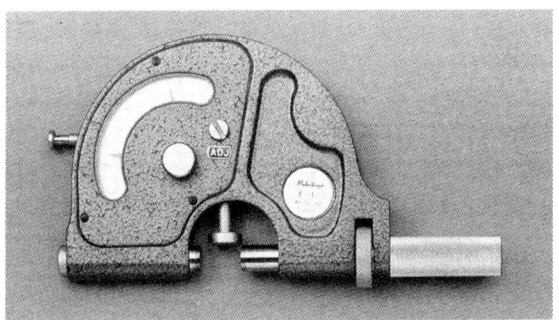

FIGURE 4.52

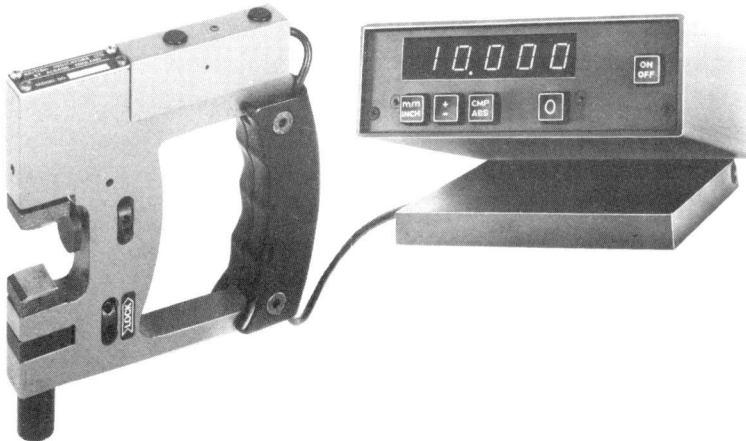

FIGURE 4.53

Chapter 5

Linear Measurements and Multi-axis Measuring Machines

In chapter 4 the point was made that, with a little ingenuity, a large proportion of measuring tasks can be undertaken with no more than a height gauge, a surface table and a dial indicator. It was also indicated that the combination can be even more versatile and easy to use if the height gauge and indicator were electronic and digital. Such electronic instruments do not use traditional machine divided and visually assessed scales, but instead have photoelectric, magnetic or capacitance devices which, in conjunction with electronic counters, can very easily record linear measurements to a high order of accuracy.

5.1 ELECTRONIC MEASURING SYSTEMS

An example of a magnetic system is shown in Fig. 5.1.

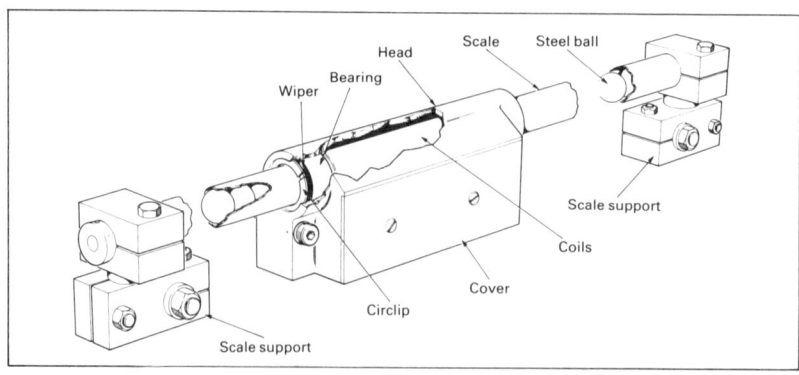

FIGURE 5.1

The master "scale" consists of accurately sized steel balls packed into a steel tube. The reading head which passes over the tube consists basically of coils which generate a magnetic field. This varies according to the position of the head relative to a ball within the tube, and measurement of the magnetic field enables the position of the head to be determined.

By the use of such refinements as preloading the balls axially within the tube, and averaging a number of readings over different balls, the accuracy of such a system can be as high as a few millionths of an inch.

Figure 5.2 shows a photoelectric system.

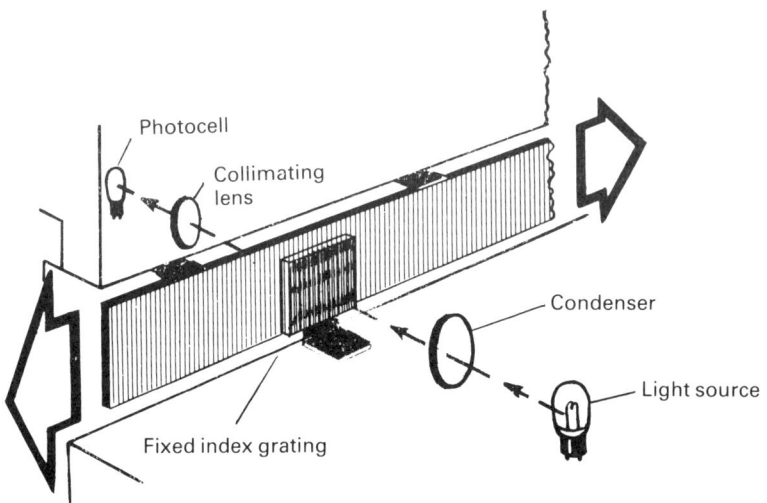

FIGURE 5.2

The basic element of this system is a grating, which is a glass scale containing on its surface a large number of accurately spaced parallel lines. In the system shown, the reading head also consists of a grating, but much shorter, and this is mounted with its lines at a small angle to the lines on the main grating. If light is viewed through both gratings mounted in this way, dark bands will be seen perpendicular to the ruled lines. If the gratings are then moved relative to each other in a direction perpendicular to the ruled lines, the bands will also appear to move but in a direction at right angles. The number of fringes to be seen and the speed with which they appear to move depends on the angle between the two sets of grating lines, and this is used to determine the resolution required.

In practical use, a light source is placed on one side of the pair of gratings and a photoelectric detector on the other. This produces electrical signals as the dark bands pass across it and these can be electronically converted into a linear dimension. Typically, the gratings are likely to have between 100 and 200 lines per millimetre, produced to an accuracy of around 0.002 mm in 250 mm, and the angle between the gratings will be such that about four dark bands are produced for a linear movement equal to the grating pitch. This will give a discrimination potential of four times the grating pitch, or up to about 0.001 mm.

5.2 MULTI-AXIS MEASURING MACHINES

Figure 5.3 shows another example of the digital electronic height gauge, already referred to in Chapter 4.

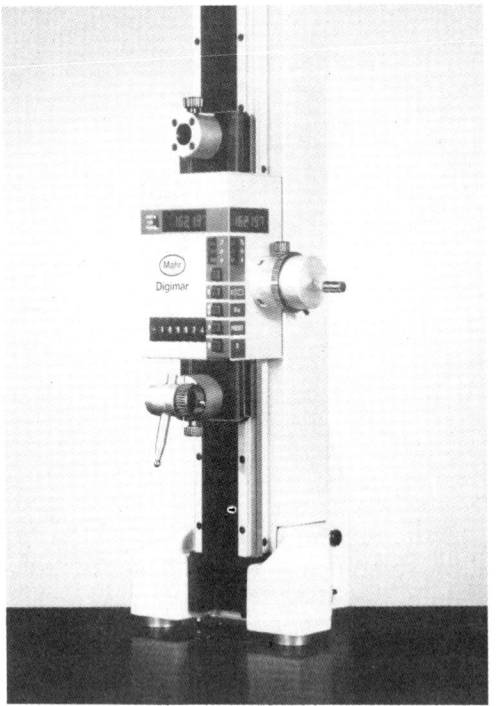

FIGURE 5.3

This example is shown with a solid ball ended measuring probe, but this again can be replaced by an electronic device which can produce an electrical signal when contact is made. The precision with which the position is defined by such a signal is remarkably high, and, given that the rest of the equipment is of a comparable standard, measurements can be reliably taken to 0.002 mm. Further developments in probe design are discussed later in this chapter, but even a simple probe used with an electronic height gauge is a powerful measuring tool which can operate very rapidly.

Such a system would be defined as a single axis system in that measurements can only be made in a direction perpendicular to the surface on which the height gauge is mounted. It is, however, not difficult to imagine the scope being considerably increased by mounting the height gauge itself against a linear scale so that its position can be measured in a horizontal direction. This gives the ability for measurement in both horizontal and vertical directions and an instrument will have been created which would be described as two axis, as illustrated in the diagram in Fig. 5.4.

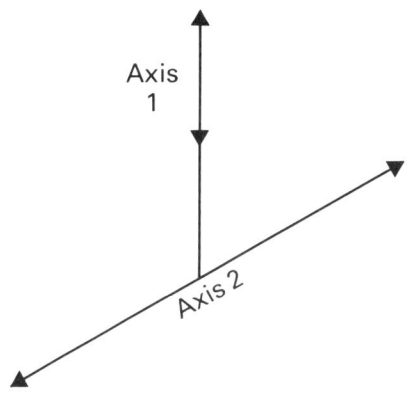

FIGURE 5.4

The next obvious step is to provide a means of moving the height gauge against a scale in a horizontal direction at right angles to the first horizontal movement, as shown diagrammatically in Fig. 5.5.

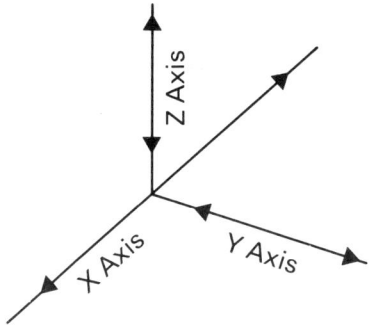

FIGURE 5.5

The instrument would then be a three-axis machine capable of measuring dimensions in any direction on a solid object.

The machine shown in Fig. 5.6 clearly illustrates these principles. In this machine, the second horizontal axis is the arm holding the probe which moves within its location on the vertical column.

FIGURE 5.6

Such logic has led to the development of the universal measuring machine which is now available in very large variety. Whilst the choice of machines can be bewildering, all operate on similar principles and can be expected to give similar performance. Selection of a suitable machine then becomes a question of determining the size of machine required, and assessing the cost of features which are particularly desirable against their potential value. Since the cost of the basic machine increases substantially as the size increases, it is important not to be tempted into purchasing a machine which is unnecessarily large. If the maximum capacity of the machine is only likely to be used on rare occasions, it might be more economical to choose something smaller, and resort to the use of first principle methods when the machine capacity is not sufficient.

Almost without exception, machines currently produced use granite as the base surface. The simplest varieties have a single X axis rail parallel to the longest dimension of the base table, and the Y axis on a cantilever which runs along this rail. The Y dimension is obtained from the position of the measuring probe along the cantilever, and the Z axis dimension is the vertical position of the measuring probe itself.

A machine of this cantilever type is shown in Fig. 5.7.

Such a machine has the advantage of being simple in principle and having, in theory, unlimited capacity in both the X and Y axis directions. In practice, however, it is limited in its Y axis measuring range by the tendency to distort due to the overhanging weight of the cantilever.

FIGURE 5.7

This problem can be overcome by supporting the cantilever at its free end, as in the machine shown in Fig. 5.8.

This change is accomplished by mounting a second X axis rail parallel to the original at surface table height, together with the necessary moving column. The support provided at both ends changes the cantilever to a bridge. This is the second, and perhaps most popular, type of arrangement. There is virtually no limit to the length of the base table and machines can be obtained with very large X axis capacity. Another advantage of this arrangement is that the bridge can be moved to one end of the machine and the table used for certain measurements in the normal and traditional way.

Earlier machines generally used steel for the essentials of the construction, but machines using light alloys, plastics and granite have been available.

64 Linear Measurements and Multi-axis Measuring Machines

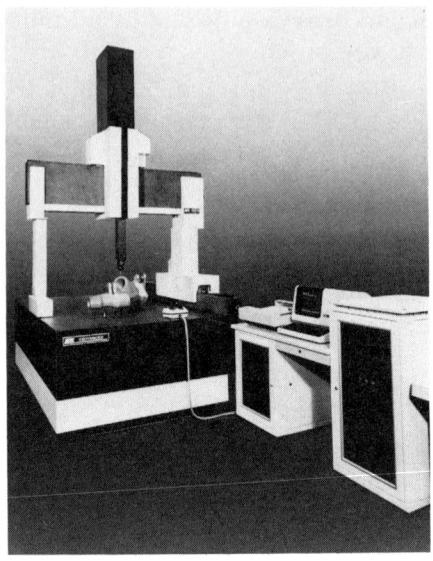

FIGURE 5.8

As a material for constructing the bridge, granite is heavy but it is extremely stable. The weight of moving parts can be overcome to some extent by the use of air bearings, the principle of which is shown diagrammatically in Fig. 5.9.

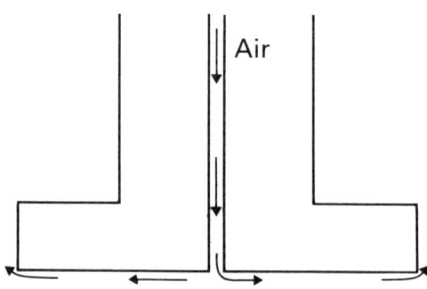

FIGURE 5.9

There are no moving parts and the members moving relative to each other are separated by a thin film or air which is pumped between them. They can operate very successfully with a gap as small as 0.1 m, and friction is almost nil.

Movement in both bridge and probe can easily be achieved manually, even on very large machines, but many are fully motor operated and, as described later in the chapter and in Chapter 12, such machines can operate automatically under the control of a prepared programme.

Although the bridge-type machine can have a very large capacity, particularly in the X axis, the machines which have the largest capacity in all directions are of four-poster construction as shown in Fig. 5.10.

FIGURE 5.10

Machines of this size are generally floor mounted and the workpiece is likely to be located on a surface table which is free standing between the posts. It is, of course, necessary for the whole of the base area to be very rigid and to be isolated from the surrounding floor surface. This is usually achieved by mounting the machine on a concrete block "floating" in sand.

At the other end of the size scale, small measuring machines are available which are suitable for mounting on a bench or surface table. Examples of such machines are shown in Fig. 5.11.

The machines so far described are defined as three-axis, X, Y and Z. If a horizontal rotating table is included, as in Fig. 5.12, a vertical axis of rotation is added to the three axes of linear movement and the machine becomes four-axis. If the facility for rotation in a vertical plane is provided, the ultimate of a five-axis machine is reached.

66 Linear Measurements and Multi-axis Measuring Machines

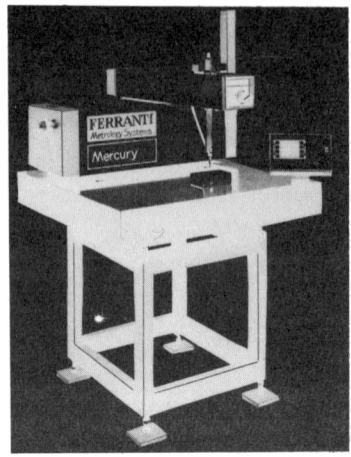

FIGURE 5.11

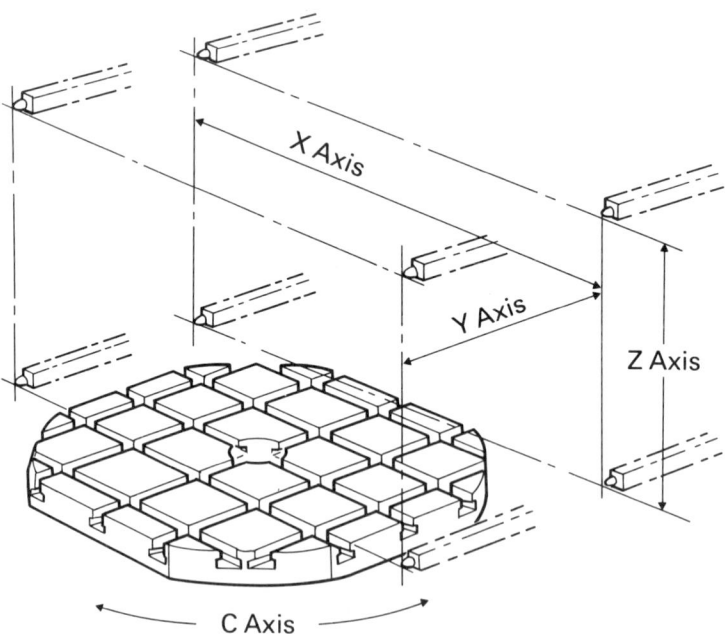

FIGURE 5.12

Whilst the range of machines available from makers' catalogues is large, many special purpose machines are made to carry out specific tasks. Such machines are frequently to be found in the automotive industry, and a machine specifically developed for the inspection of a vehicle body is shown in Fig. 5.13.

Multi-axis Measuring Machines

FIGURE 5.13

The success of multi-axis, co-ordinate measuring machines is largely due the development of accurate, but comparatively inexpensive, measuring systems giving their output as electrical signals. These signals are fed into computing devices which not only assess the dimensions of the part being measured, but also provide a variety of analytical statistics. The use of a computer in this way is dealt with again in Chapter 12, but the simplest use of the electrical output is to provide either a direct digital or analogue reading, or a signal indicating whether the dimension is satisfactory or not.

If the measuring machine has motor operated movements, its operation can be programmed to carry out the inspection of a complex part quite automatically. Programmes can be prewritten, but the capability of using the manual setting of a first component as a means of obtaining the programme is readily available.

The output from the machine can also be fed to a printer which will give hard copy of all or selected dimensions. The printout can be programmed to carry the identity of the part being measured and such details as date and identity of operator.

Although machines having all these facilities are inevitably expensive, they can carry out much more satisfactorily the tasks performed by the traditional receiver gauges mentioned in Chapter 2. For small batches of work they are invaluable, and even for large

68 Linear Measurements and Multi-axis Measuring Machines

quantities the economics are such that it is rarely justifiable to use a single purpose receiver gauge.

A particularly sophisticated multi-axis measuring machine is shown in Fig. 5.14.

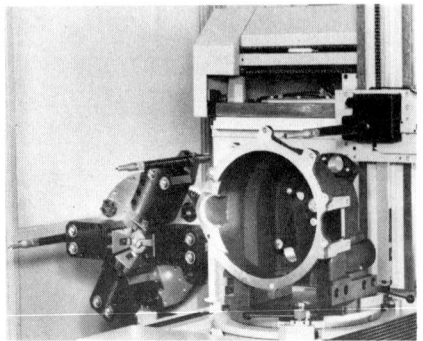

FIGURE 5.14

This machine not only has automatic programmed movement in four-axis, but also has a magazine containing a variety of measuring probes which are automatically changed during the measuring cycle.

5.3 MEASURING PROBES

It will be obvious that a vital element in the measuring machine is the means by which the surface to be measured is contacted. In the measuring examples of the last chapter, a dial indicator or its electronic counterpart was used for this purpose.

The dial indicator, particularly if of the cantilever type, is still capable of being a simple and useful probe, but other devices can be more appropriate in the measuring machine context.

The movement, particularly of small machines, is free enough for rigid probes, such as those shown in Fig. 5.15, to be useful.

For the measurement of dimensions to flat surfaces, the probe may be hand held against the surface to be measured. A ball ended probe is perhaps the most suitable for use in this way.

Rigid probes with conical ends are used for measuring hole positions. Again, the freedom of movement of small machines will allow the axis of a conical probe dropped into a hole to take up the position of its centre.

Measuring Probes 69

FIGURE 5.15

The versatility of multi-axis measuring machines has, however, been considerably enhanced by the development of contact probes, an example of which is shown in Fig. 5.16.

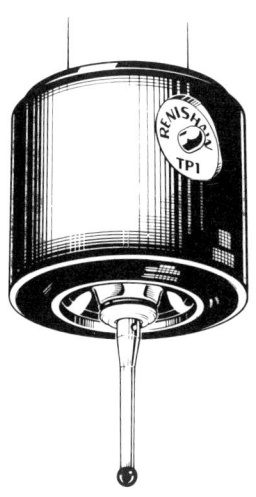

FIGURE 5.16

This type of probe is available in two forms. It may only indicate a contact position, or it may also have a small measuring capability. In the analogue measuring mode, such a probe has an accuracy of 0.5% of its reading over a range of 1 mm in each of five directions. These measured movements, as shown in the diagram in Fig. 5.17, are in either direction in two horizontal planes at right angles, and in a positive direction only in a vertical plane.

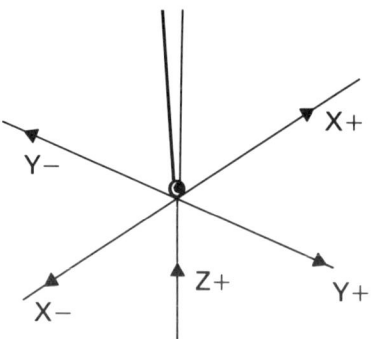

FIGURE 5.17

The measuring capability of the probe is perhaps most useful for in-process gauging, as described in Chapter 14. For measuring machines, a probe registering only the contact is likely to be the most convenient and to give the most consistently accurate results. This probe is known as a "touch trigger" probe, since it has no measuring capability, but triggers an electrical signal at the touch position. The main features of the internal construction are shown in Fig. 5.18.

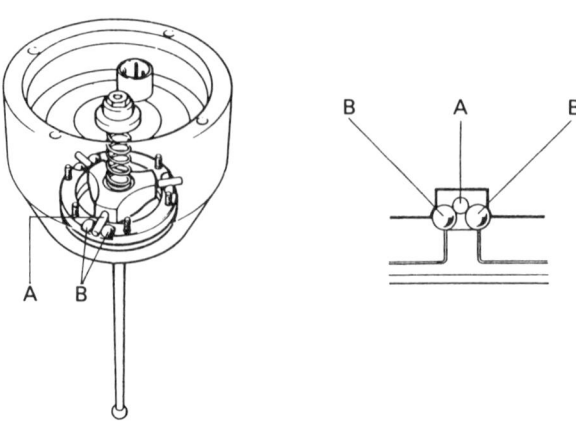

FIGURE 5.18

It will be seen that the stylus is rigidly mounted in a carriage having three projecting rollers A. This assembly is located by a spring against three pairs of balls B mounted in the probe head. All six balls and the three rollers are linked to form an electrical circuit which is disturbed by any movement of the stylus relative to the head. As a simple example of the use of such a probe, if it is required to measure a dimension X in Fig. 5.19, the probe is moved until an indication that face A is in contact is obtained.

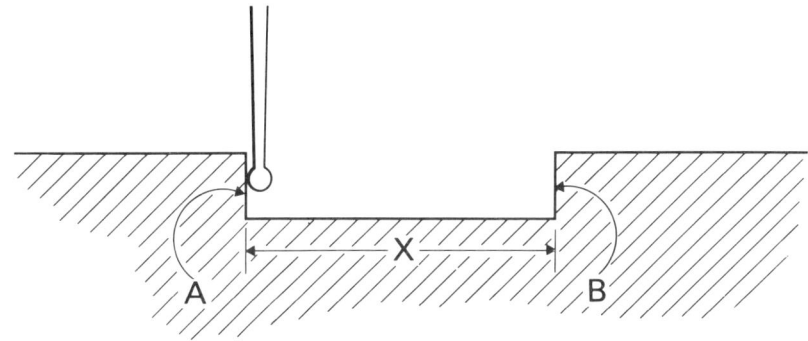

FIGURE 5.19

The measuring machine reading is then taken, and the machine head is moved until a contact indication is obtained from face B. The dimension required is then the difference between the two readings less the diameter of the probe stylus.

In practice, this operation is likely to be rather more sophisticated. The machine movements will probably be motorised, and the signal from the probe will cause the machine reading at that point to be recorded automatically, and it will also cause the machine movement to change direction. This makes possible fully automatic inspection of a complete component, as described in Chapter 12.

The accuracy of this form of measurement depends partly on the measuring machine used, but the probe indications are repeatable to within 0.1 μm.

To facilitate the use of the probe in difficult positions, a large variety of styli are available, some of which are shown in Fig. 5.20. Since the probe only indicates contact, there are no problems of allowing for the length and angle of the stylus.

The versatility of the probes can also be increased by the use of mountings which can hold more than one probe, as shown in Fig. 5.21.

72 Linear Measurements and Multi-axis Measuring Machines

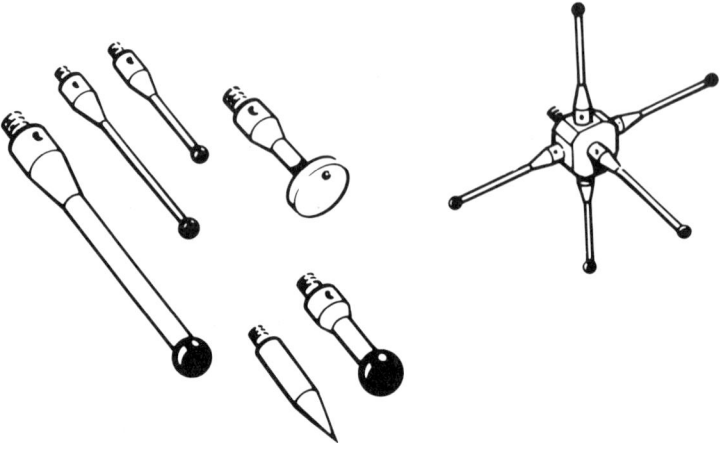

FIGURE 5.20

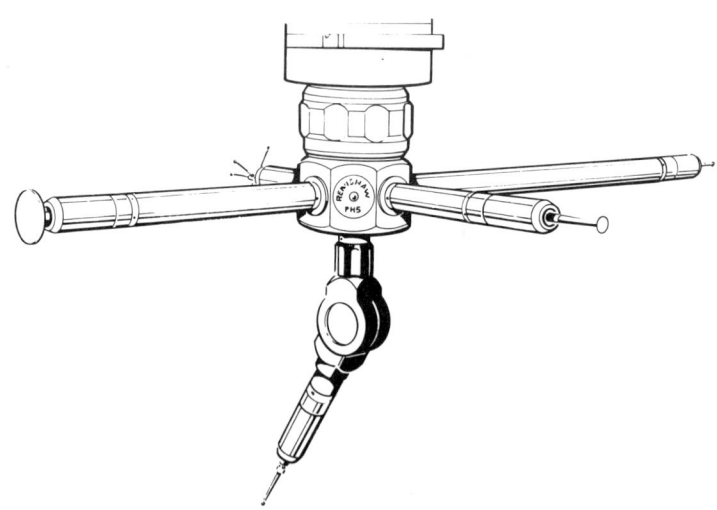

FIGURE 5.21

A further development is the motorised probe head shown in Fig. 5.22.

This head contains a special type of motor which can move the probe to a large number of predetermined positions with a repeatability of 0.5 μm. In addition to avoiding the need for

Measuring Probes 73

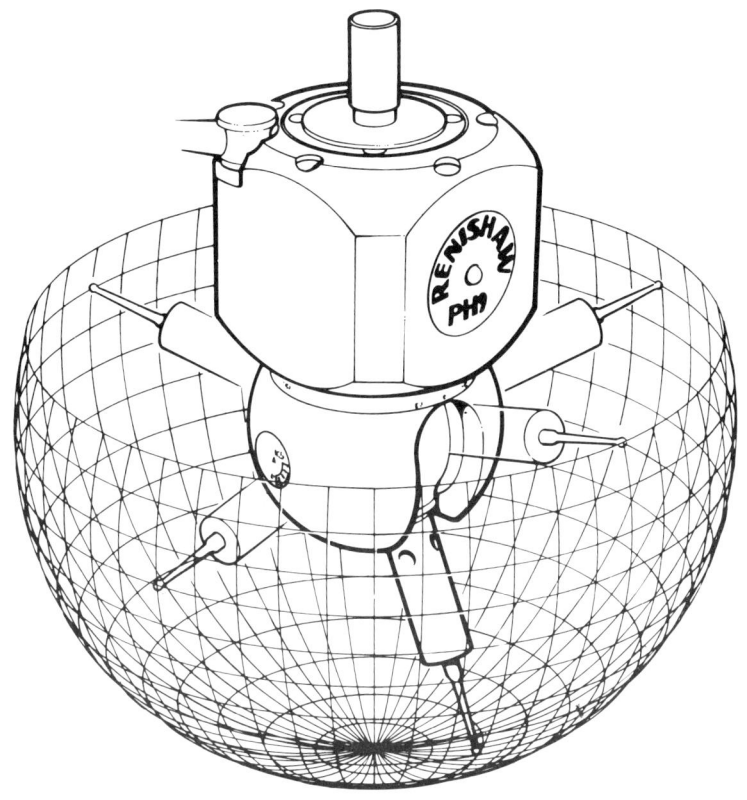

Figure 5.22

complex stylus assemblies, the movements of the probe can be computer controlled and programmed, making it possible for the probe head to carry out movements which would otherwise have to be made by the machine itself.

Chapter 6

Angular Measurement

In order to specify an engineering product fully, several different forms of dimension may be required. The majority will probably be straightforward linear dimensions, but it might also be necessary to specify angle and angular relationships. Such relationships may take a variety of forms; for example, between two surfaces, as in Fig. 6.1, or between two holes, as in Fig. 6.2.

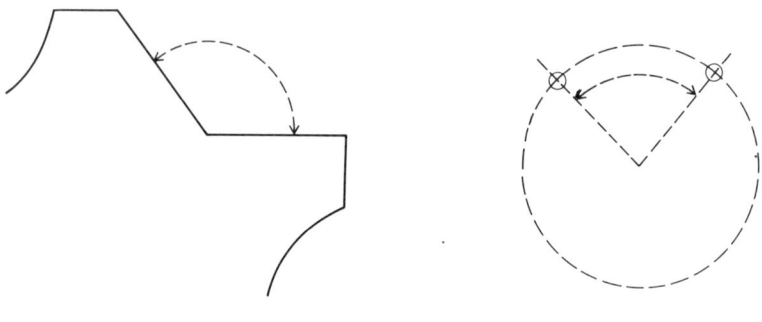

FIGURE 6.1   FIGURE 6.2

6.1 SPECIFICATION OF ANGLES

It is first necessary to consider the units in which the angle is specified. The basic unit is almost certain to be degrees. Radians is an alternative, but is rare in engineering work. The degree is likely to be subdivided into minutes and seconds of arc, but there is a trend towards subdivision into decimals of a degree. Some confusion will result if the method of subdivision is misinterpreted. For example, an angle of 25.50 degrees is 20 minutes smaller than an angle of 25 degrees 50 minutes.

Specification of Angles 75

In the case of rotating and circular parts, angles are commonly used to specify hole positions. This might be directly in terms of the angle between one hole and the next, but it might also be expressed as a number of holes equally spaced, as shown in Fig. 6.3.

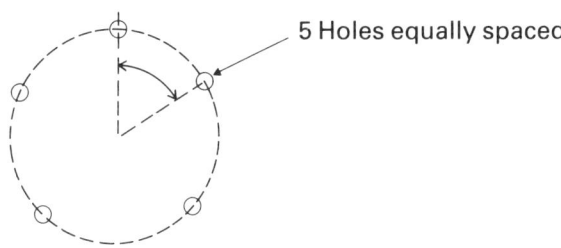

FIGURE 6.3

In this situation it is a simple matter to calculate the angle between the holes, i.e. $360° \div 5 = 72°$.

For high-speed rotating assemblies, where balance errors caused by reassembly of parts in different positions can be a problem, it is common practice to specify all but one of the holes as equally spaced, as the example in Fig. 6.4 shows.

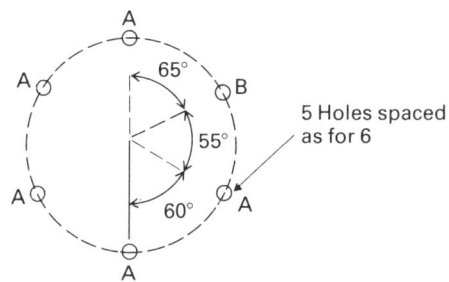

FIGURE 6.4

This illustration shows five equally spaced holes A spaced as for six, i.e. 60° apart. The sixth hole B may then be individually dimensioned anywhere within an arc of 120°, perhaps 65° from one hole and 55° from the other. This technique ensures that the assembly can only be built in one position.

6.2 USE OF LINEAR CONVERSION

Although there is no shortage of equipment for the measurement of angles directly, such equipment might not be readily available and, for a comparable order of accuracy, it is likely to be more expensive than equipment for linear measurement. It is, therefore, often convenient and economical to use a little simple trigonometry and convert an angle to linear dimensions. In the example of Fig. 6.5, the position of hole 1 relative to hole 2 can be converted from an angle of 30° to two linear dimensions, X and Y, at right angles to each other.

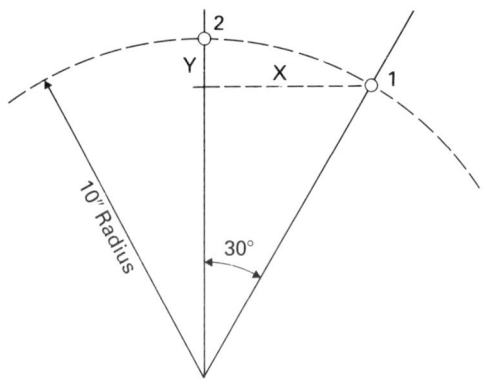

FIGURE 6.5

Dimension X will be $10 \times \sin 30° = 5''$, and dimension Y will be $10 \times \cos 30° = 3.4''$. This technique is widely used and is particularly valuable if high accuracy is important.

6.3 THE PROTRACTOR

The simplest instrument for the measurement of angles is the protractor. Its basic form is the angular equivalent of the steel rule, merely a flat piece of material marked with angular divisions, as shown in Fig. 6.6.

In this form, it is of little value to the engineer and perhaps the simplest practical form is of the type used in conjunction with a steel scale shown in Fig. 6.7.

The Protractor

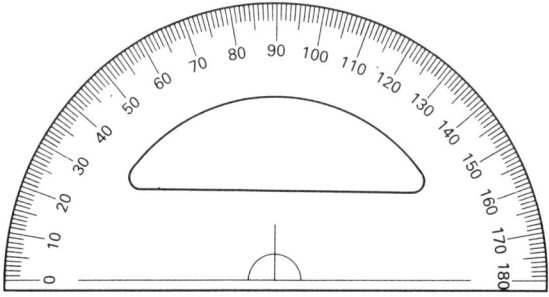

FIGURE 6.6

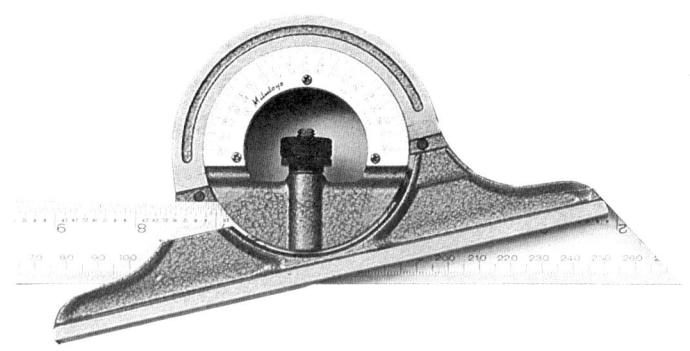

FIGURE 6.7

The steel scale in this application contains a machined groove into which the protractor is located by a tongue. Within the protractor itself, the tongue is rigidly attached to an angular scale, whilst the base surface may rotate relative to it. The position of an index mark indicates the angle between the protractor's base surface and an edge of the scale.

The accuracy of this type of instrument is low, not much better than ± 1°, and a more accurate, and very popular, version is shown in Fig. 6.8.

This instrument is known as a bevel protractor and, whilst its principles are similar to the protractor shown in Fig. 6.7, its method of construction ensures greater accuracy. It is often fitted with a vernier scale, as the example in Fig. 6.8, and in this form it is capable of accuracy of about 10 minutes of arc. The vernier scale for

78 Angular Measurement

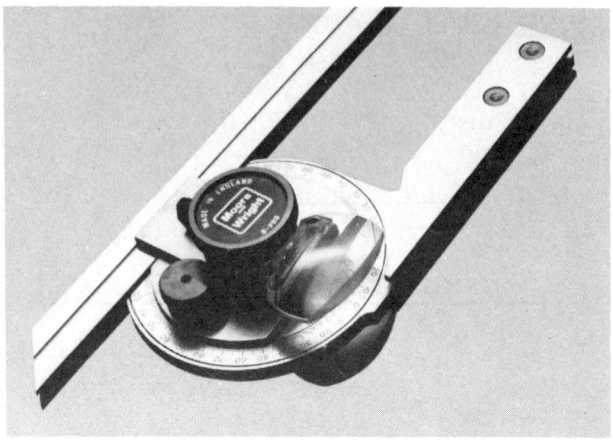

Figure 6.8

angular measurement is produced on the same principle as the linear vernier scale described in Chapter 4.

Another version of the protractor, which gives the reading on a dial indicator, is shown in Fig. 6.9.

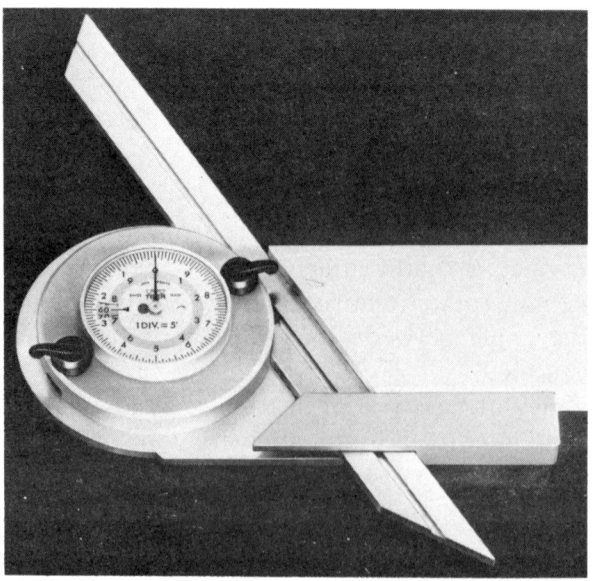

Figure 6.9

6.4 THE SPIRIT LEVEL

Although the spirit level might not normally be regarded as an angular measuring instrument, in fact it measures the angle between a given surface and the horizontal. With very little adaptation it can measure any angle provided that it is in a vertical plane.

Figure 6.10 shows a simple but very accurate level.

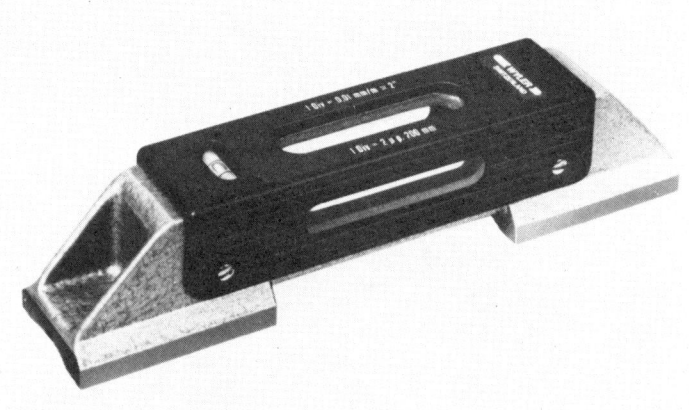

FIGURE 6.10

The accuracy depends on the curvature of the tube containing the bubble, but it is not exceptional for a level such as this to be able to be read consistently to 4 seconds of arc.

The spirit level shown in Fig. 6.11 is mounted on a circular base which rotates within an accurately produced square frame.

FIGURE 6.11

80 Angular Measurement

This enables the angle of any surface to be measured with respect to the horizontal or vertical, but it must be remembered that, in this application, the accuracy is likely to be controlled by the accuracy with which the angle between the circular base and the square frame can be read. In this example, a vernier scale allows the angle to be read to 3 minutes of arc.

A more accurate version of this instrument, known as a Clinometer, is shown in Fig. 6.12.

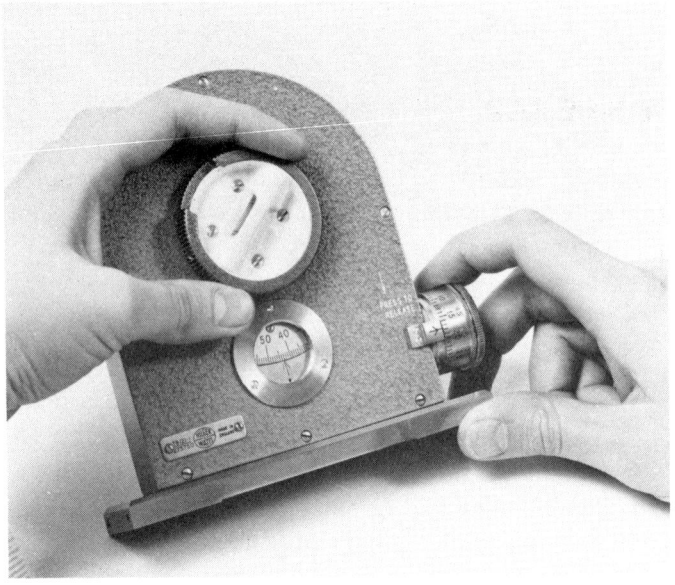

FIGURE 6.12

In this form, the movement between the spirit level mount and the external casing is controlled by a micrometer screw, which increases the accuracy to about 1 minute of arc.

Electronic varieties of these instruments are available, and these make use of a pendulum rather than a bubble as the datum. The position of the pendulum is electrically sensed and the readout can be in digital form, as the example in Fig. 6.13 shows.

Such an instrument can display the angle in units as small as 0.2 seconds of arc.

FIGURE 6.13

6.5 THE SINE BAR

Very accurate setting up of angles can be achieved using a completely different technique and a device known as a sine bar. A sine bar is shown in Fig. 6.14, and it will be seen that it consists simply of a flat rectangular section to which two rollers are rigidly attached.

The distance between the rollers is very important and determines the accuracy of the instrument. This distance is usually in multiples of 5 inches or centimetres for ease of calculation. The instrument

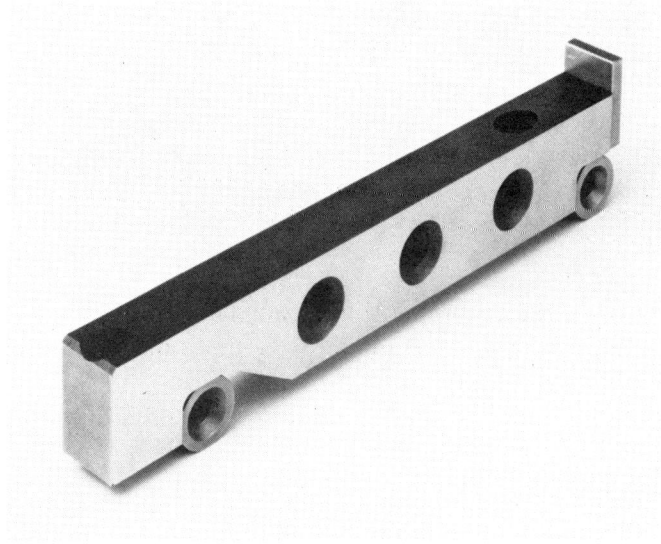

FIGURE 6.14

82 Angular Measurement

obtains its name from its use as the hypotenuse of a right-angled triangle in which the subject angle is subtended by this hypotenuse and the base on which it is placed. As shown in Fig. 6.15, the perpendicular dimension of the triangle is then formed by gauge blocks of such a length that division by the length of the sine bar gives the sine of the required angle.

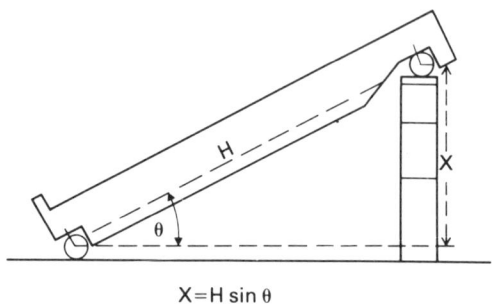

$X = H \sin \theta$

FIGURE 6.15

This method is difficult to use as a direct means of measurement, but it can be a very accurate angular comparator.

A useful development of the sine bar is the sine table, as shown in Fig. 6.16.

FIGURE 6.16

This enables a plane, on which the workpiece can be mounted, to be set up at an angle to the surface on which the sine table is placed.

An obvious further development is the compound sine table, which is effectively two sine tables mounted at right angles on top of each other. The particular value of this device, shown in Fig. 6.17, is that it is possible to mount the most complex part on the table in such a way that any chosen surface is parallel to the base surface.

FIGURE 6.17

6.6 ROTATING TABLES

The most popular machine for the measurement of angles is the rotating table. A selection of rotating tables is shown in Fig. 6.18 and this gives an indication of the variety available.

In principle, the rotating table is very simple, and, in addition to its angular measurement capability, it is a valuable piece of equipment for checking roundness and concentricity. The cost of a table in its simplest form will be determined by the accuracy of its rotation and the sophistication of its measuring system. As with other measuring devices, in the interests of economy, it is important that the accuracy of the table is not excessive in comparison with the tolerances of the work being measured.

84 **Angular Measurement**

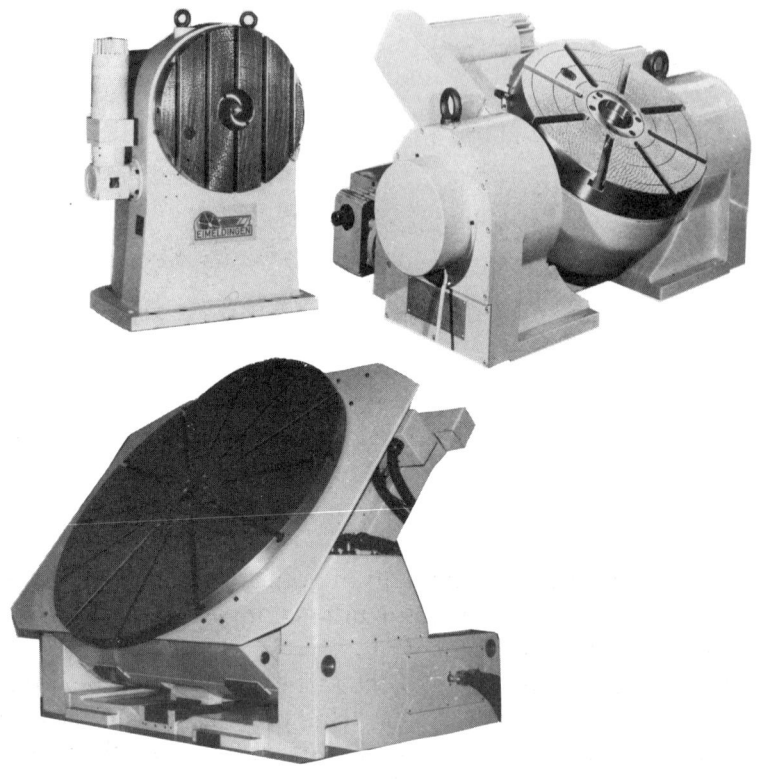

Figure 6.18

The tables may be rotated directly by hand through some form of gearing, or they may be motorised. Small tables often have a worm gear rotating mechanism and the rotation of this gear can be calibrated to read the angular position of the table, as the example in Fig. 6.19.

This system has a weakness in principle in that the position of the table is being measured indirectly, and the readings taken will be affected by any inaccuracies in the gear train.

It is fundamentally better for any measurements to be taken directly from the table itself, and many tables have angle scales built into their rim. Obviously, the accuracy of such a scale will be proportional to the diameter of the table, but, whatever the diameter, the accuracy will again be enhanced by the provision of a vernier scale.

Rotating Tables 85

Figure 6.19

However, with the development of optical and electrical measuring devices, most tables used for inspection work have measuring systems which are not mechanical.

Optical systems generally project part of a circular scale attached to the table on to a small screen. This process will, in itself, give some magnification. The accuracy of reading is then further improved by means of an optical vernier. An example of the output from such a system is shown in Fig. 6.20.

Figure 6.20

86 Angular Measurement

The upper scale is graduated in degrees and tens of minutes and each ten-minute interval is defined by a pair of parallel lines. This scale is projected against a single line graticule rigidly linked to the lower vernier scale, and both graticule and vernier scale are moved by a vernier hand wheel. In operation, the vernier wheel is rotated until the single line comes between a pair of ten-minute lines. Assessment of a central position can be done by eye with surprising accuracy and readings can be taken consistently to 1 second of arc. In the example of Fig. 6.20, the angle from the upper scale is 120 20' whilst the vernier scale reads 3' 16", giving a total reading of 120 23' 16".

Whilst such optical systems are very accurate and reasonably easy to use, they are expensive, and with the rapid development of electronic measurement and display systems the most attractive means of displaying the angular position is a digital system. It is not only the simplest and most versatile to use, but it is likely to become much cheaper more quickly than traditional optical methods.

As in the case of digital readout systems used on linear measuring machines (see Chapter 5), signals are generated within the machine which are counted electronically. Again, as in the case of linear machines, the generating mechanism can be either electromagnetic or optical. Most of these systems are capable of an accuracy of 1 second of arc, and they have the other usual electronic system advantages. An example of a table with a digital readout device is shown in Fig. 6.21.

FIGURE 6.21

Rotating tables are often used just for rotating a part on a particular axis. This enables measurements of eccentricity and roundness to be made, but for these purposes there is no need for any measurement of the angular position of the table. It is remarkably common for tables with very sophisticated measuring systems to be used primarily in this way, but it is, of course, an extravagant use of such equipment.

The need to carry out simple concentricity and roundness checks can be more economically satisfied by using a so-called "spin table", which is generally similar in appearance to the rotating tables already described but without the measuring mechanism.

An ingenious "table" aimed primarily at satisfying this requirement is illustrated in Fig. 6.22.

Figure 6.22

This, in fact, is not a complete table, but three arms attached to a central hub. This is mounted on a spindle which passes through a hole in a cast iron or granite surface table. The central hub incorporates an air bearing pad, and other air bearing pads, which may be moved to any position, are mounted on the three arms.

In use, the pads are positioned so that they support the work being measured which will then rotate very freely and very accurately on the air bearings. The whole assembly may be driven

88 Angular Measurement

by a motor mounted beneath the surface table and, if required, an angular measuring system can also be provided. The complete equipment, including motor and angular measuring device, is shown in Fig. 6.23.

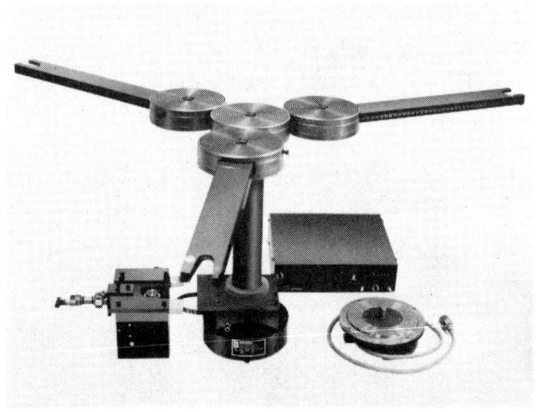

FIGURE 6.23

Although rotating tables are generally used with the axis of rotation vertical, it can be convenient for many applications if the axis is horizontal, as in the table shown in Fig. 6.21. In fact, such a rotating table mounted on a surface table can, when used with the height measuring devices described in Chapter 4, form an economical and surprisingly comprehensive measuring machine, as shown in Fig. 6.24.

FIGURE 6.24

6.7 ROUNDNESS

On certain engineering components, particularly associated with all types of bearings, there may be a requirement to measure the various parameters peculiar to cylindrical and spherical surfaces to a very high order of accuracy. For this work a wide variety of "roundness" testing machines are available. The feature of such machines, two of which are shown in Fig. 6.25, is exceptional accuracy of rotation, up to around 0.05 μm.

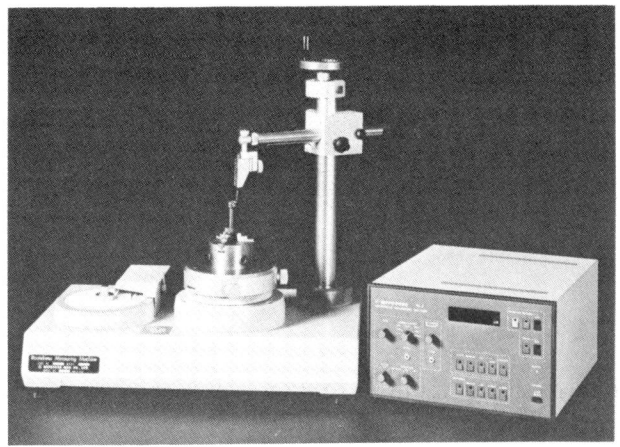

Figure 6.25

Most of these instruments are capable of giving a printed readout, often in the form of a circular graph with a large radial magnification, as shown in Fig. 6.26.

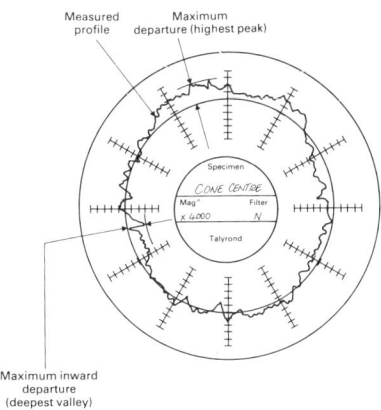

Figure 6.26

90 Angular Measurement

It must be noted when interpreting graphical representations of this type, that the large magnification in one axis only gives a very misleading visual impression. An indication of the magnitude of this distortion is that in the example of Fig. 6.26, if the diameter itself were drawn to the same magnification as the surface irregularities, it would be more than 200 metres. The magnification of 4000 in this example is quite modest for this type of work. Machines are available which will give magnifications of up to 200,000.

Chapter 7

Flatness and Surface Finish

In addition to the more straightforward dimensional characteristics of an engineering product, its performance, appearance and cost are likely to be strongly influenced by the quality of the finish on the various surfaces. This may be important for a variety of reasons. The most obvious is that the surface has a function which involves contact with another surface. This can be moving contact, as in the case of a bearing diameter, or static, as in the case of two surfaces required to provide an oiltight joint. Finish might also be important in the interests of reducing stress on the part and, particularly if on an external surface, it might be important merely for aesthetic appearance.

7.1 QUANTIFICATION OF SURFACE FINISH

Surface finish can be accurately quantified, and several different principles have been used to achieve the desirable objective of expressing the requirement and the measurement in terms of only one number. The four main methods are indicated in the diagrammatic representation of the cross-section of a surface shown in Fig. 7.1. These are Centre Line Average, referred to as R_a, Root Mean Square, referred to as RMS, maximum peak to valley, referred to as R_t, and maximum peak to mean, referred to as R_p.

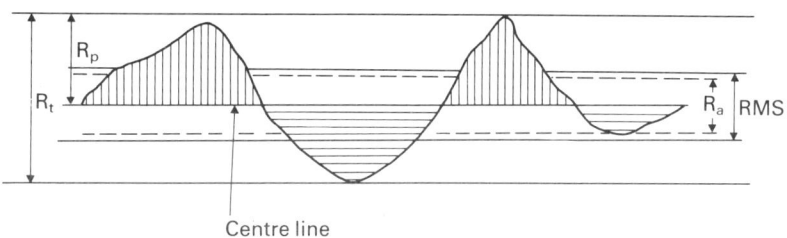

FIGURE 7.1

92 Flatness and Surface Finish

Whilst all these parameters have some relevance, depending on the role the surface has to play, the most common method is the Centre Line Average, and it is worth describing in some detail the way in which this figure is derived.

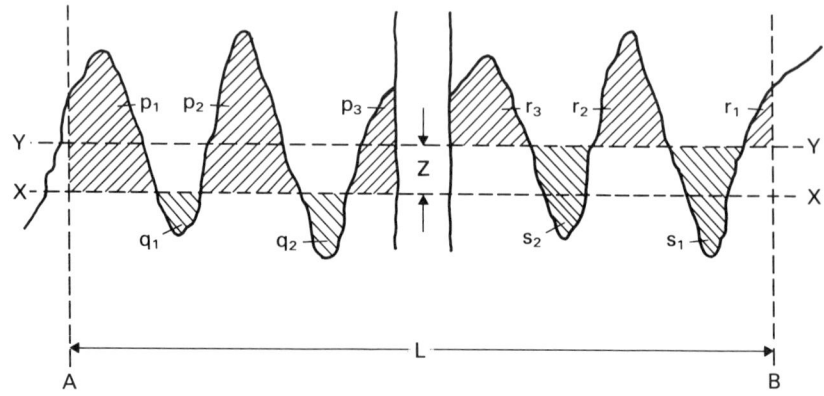

FIGURE 7.2

In the diagram at Fig. 7.2 a straight line $x-x$ is drawn by eye following the general direction of the profile and covering the sampling length L. The areas of the profile p above, and q below this line are then measured and a distance z is obtained by dividing the difference between these areas by the sampling length, i.e.

$$z = \frac{\text{areas } p - \text{areas } q}{L}.$$

If a new line $y-y$ is now drawn parallel to $x-x$ at a distance z from it, this new line will be a mean centreline, and the CLA value (R_a) will be the sum of the areas above and below this line divided by the sampling length, i.e.

$$R_a = \frac{\text{areas } r + \text{areas } s}{L}.$$

The length over which the sample is taken is obviously very important, and in the example shown in Fig. 7.2 the sample length L is adequate to give a measure of the total surface because it covers a significant number of complete surface finish cycles. If, however, the combination of sampling length and finish characteristics is such that the sample contains say, only one or even less total cycles, the result will not include all the characteristics of the surface.

The sampling length is used extensively in surface measurement to segregate the various characteristics of the surface.

As shown in Fig. 7.3, a normal finish consists of several different elements. These are referred to in this drawing as primary texture, secondary texture and form errors, but they might also be described as roughness, waviness and flatness respectively. The term "surface finish" is normally used to describe the first, and perhaps the second of these elements. Errors of flatness are usually considered, and measured, separately.

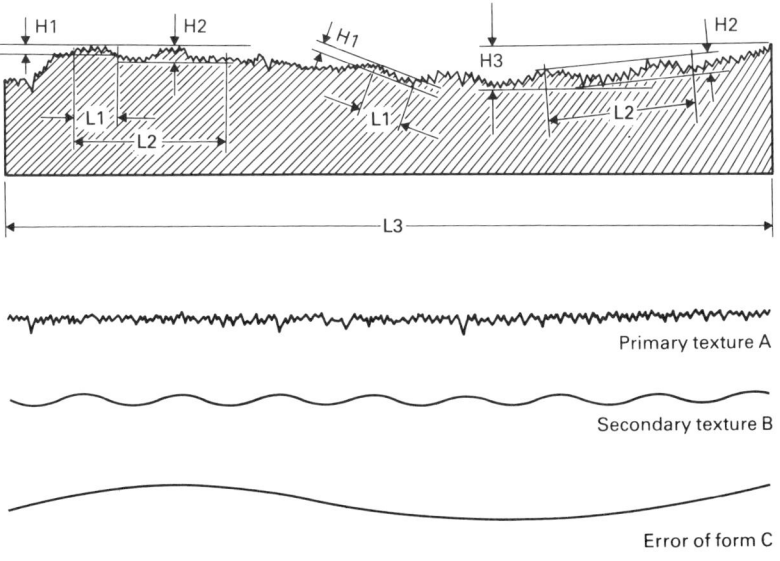

FIGURE 7.3

The parameter which is necessary to achieve this is the cut-off value. This is the length of the surface sample to be considered, and all features within this length will be included to arrive at, for example, the CLA value.

In Fig. 7.3 the effect of three cut-off values, L1, L2 and L3, is indicated. It will be seen that if a cut-off value of L1 were selected, the reading H1 obtained would cover only the primary texture. The values H2 and H3 obtained from L2 and L3, however, would, in addition, include the secondary texture and the form respectively. Whilst it is possible to obtain a finish reading for any cut-off value, there are again preferred values. The Imperial units range, for example, covers six options between 0.003 in and 1.0 in.

Since surfaces for which measurement of the finish in quantitative terms is required are likely to be fine, small sampling lengths are appropriate and the normal standard is 0.030 in. Such a sample is not likely to cover all the characteristics of a surface, but experience has shown that, in practice, it is the most useful. In metric units the figure used is 0.8 mm.

Surfaces produced by normal engineering methods would, if looked at in cross-section, generally show a different profile in different directions. Cutting processes, such as turning and boring, produce a surface which is evenly spaced and unidirectional, as indicated in Fig. 7.4.

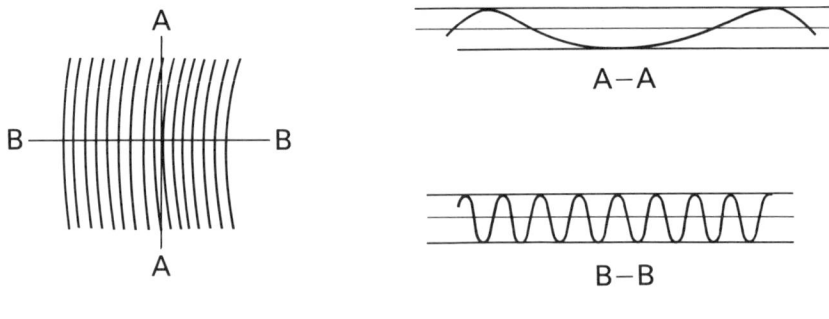

FIGURE 7.4

Grinding generally produces a surface which is unidirectional, but does not have regular cycles.

Operations such as lapping and polishing produce very fine surfaces but they are both multidirectional and irregular.

The direction in which the cutting tool moves is known as the "lay". Normally, surface finish would be measured in the direction which gives maximum roughness, and this is likely to be in a direction at right angles to the lay.

7.2 SURFACE FINISH UNITS

The metric unit in which surface finish is normally measured is the micron which is 10^{-3} mm. It is still, however, quite common to measure surface finish in Imperial units and in this case it is expressed in micro-inches denoted by μ'' (or μ in). The micro-inch is 10^{-6} in, hence 1 micron is equal to 40 micro-inches.

In spite of the fact that surface finish is measured in very small units, its measurement involves a fair amount of approximation, but experience has shown that such approximations are very satisfactory for engineering applications.

In practice, both the function and the appearance of a surface are proportional to a geometric progression of the numerical value and the requirements are generally defined in terms of a scale of preferred values structured in this way. The preferred values for Imperial units, for example, are 1, 2, 4, 8, 16, 32, 64, 125, 250, 500 and 1000 micro-inches.

7.3 MEASUREMENT OF SURFACE FINISH

The simplest methods of assessing the finish, which are quite appropriate in many cases, are subjective and make use of the senses of sight and touch. With experience, eyes and fingernails can often be adequate assessors of a surface.

A degree of sophistication can be introduced into these methods by the use of comparative standards, as shown in Fig. 7.5.

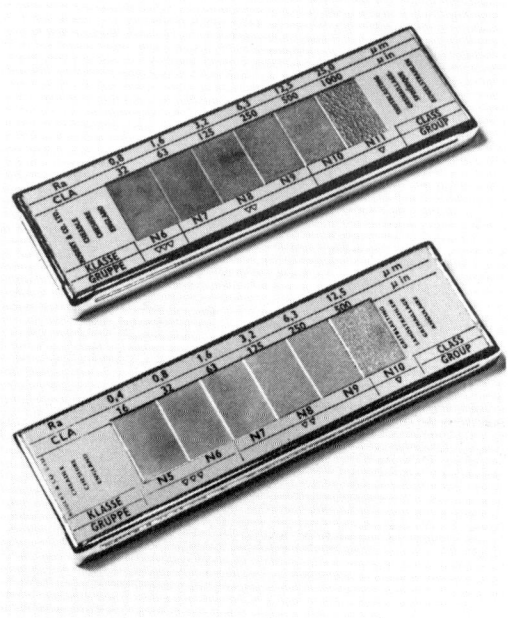

FIGURE 7.5

These standards are calibrated samples which are compared, by sight and touch, with the surface to be measured. The sets contain samples graded through the range of preferred values and it is not too difficult to assess the finish being measured relative to two adjacent samples. For maximum sensitivity using this method, the samples should be of a similar material and produced by the same machining process as the surface being assessed.

7.4 SURFACE FINISH MEASURING MACHINES

For more accurate assessment of the surface, many instruments are available which will quantify the finish. Most of these move a pick-up over the surface to be measured and will immediately give a direct reading of the CLA value. Partly because of the very small units of measurement involved, the readings will appear to be a very precise, but, because they are taken over a very small sample of the surface, the figures will only be an indication of the finish of the surface as a whole.

The principle of most surface finish measuring devices is to obtain an electrical output from the movement of a diamond stylus, similar to that in a record player, as it traverses across the surface. As shown in Fig. 7.6, the stylus is mounted in a head which is supported on a shoe or skid by the surface being measured. A skid is a rigid support with a curved under surface, whilst a shoe is a support which is flat but pivoted.

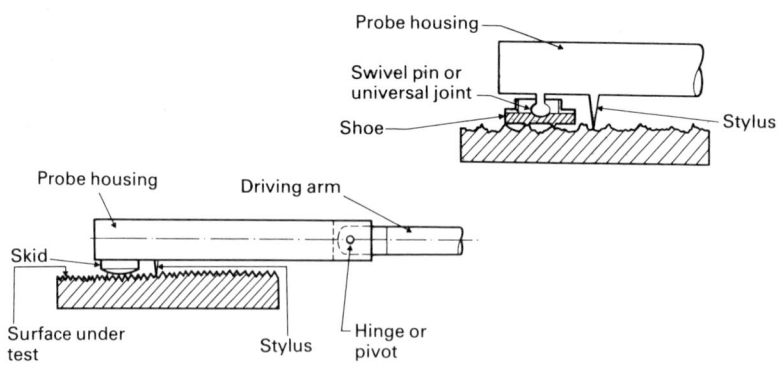

FIGURE 7.6

The head assembly is free to move vertically, and its output is processed electronically to give the appropriate surface finish figure.

Some instruments are available in which the stylus assembly is moved across the surface by hand. This is not a particularly satisfactory method and in the majority of machines this movement is motorised.

The amount of traverse required to obtain a reading is normally dependent on the cut-off value chosen. Obviously, a more representative reading will be obtained if the mean of several checks is taken, and it is usual for the traverse to be a little over five times the cut-off value, so that the reading offered is the mean of five successive checks.

In addition to giving a single quantified figure for the quality of the surface, many instruments will also produce a graphical recording of the type shown in Fig. 7.7. Such a recording is valuable for assisting process development and for basic engineering work.

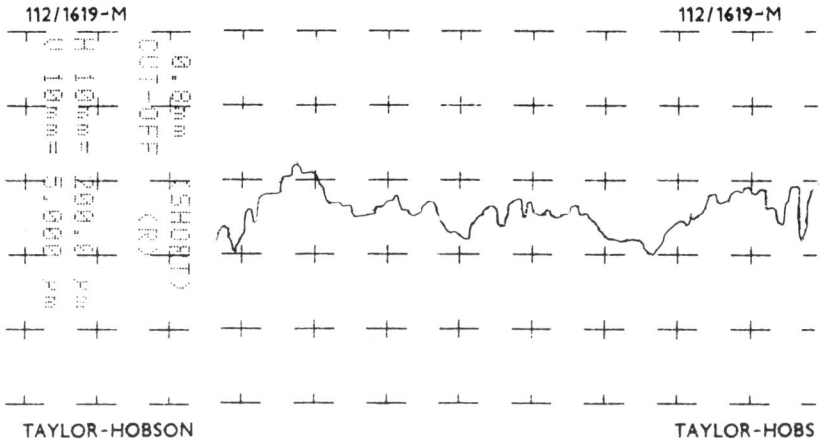

FIGURE 7.7

It must be remembered, however, that, as in the case of the roundness recording shown in Chapter 6, this graphical representation does not give a magnified picture of the surface, since the vertical magnification will be several orders of magnitude greater than the horizontal magnification. In the example of Fig. 7.7, the vertical scale is magnified 40 times more than the horizontal, and to obtain a true magnified picture, the graph would have to be stretched horizontally to a length of more than 3 m.

Whilst graphical recordings are not dependent on such parameters as sampling lengths, their shape will be determined to some extent by the geometry of the measuring probe and the way it is supported whilst traversing the surface.

An important factor, whatever the form of output, is the shape of the stylus. The normal geometry is a 60° or 90° cone with an end radius of between 0.0001″ and 0.0005″. Whilst such a stylus might not penetrate to the bottom of the finest scratches, this has no significant impact on the validity of the results.

The other factor of the instrument which will influence the reading is the relationship between the stylus and the shoe or skid. Since the stylus and the shoe do not make contact with the surface at the same point, the reading obtained will be affected by the positions they take up relative to the surface irregularities. Figure 7.8 illustrates diagrammatically the extreme situations.

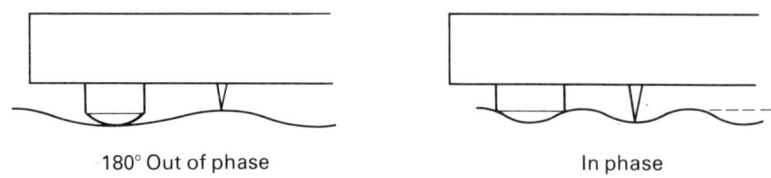

FIGURE 7.8

If the finish is cyclic and of an amplitude equal to the distance between stylus and shoe, they will move vertically together and no reading will be obtained. On the other hand, if the finish cycle and the head dimension are 180° out of phase, the readings will be doubled. However, experience again shows that for measurements on fine finishes this does not detract from the value of the readings obtained as an indication of the potential performance of the surface.

The machines themselves are available in many levels of sophistication, but the choice is basically between an instrument which is capable of comprehensive analysis of a surface, and one which goes a stage beyond assessment by sight and touch in giving a numerically quantified reading on the shop floor.

An example of one of the more versatile machines is shown in Fig. 7.9.

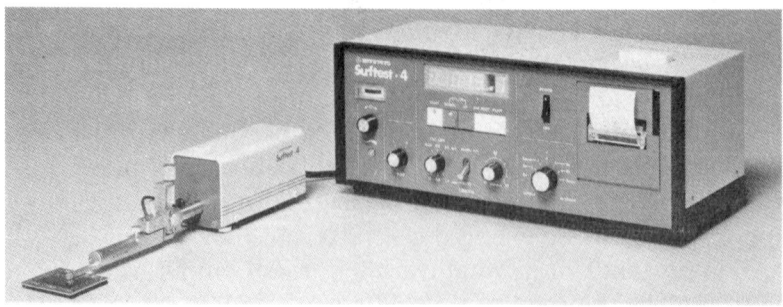

FIGURE 7.9

Surface Finish Measuring Machines 99

Perhaps the most generally useful machines are those which are produced in modular style. An example is shown in Fig. 7.10 and the advantage is that only sufficient modules need be purchased to carry out the requirements of the time, which may be no more than a simple CLA figure on flat external surfaces.

If at a later date it is desired to add graphical recording, ability to measure more complex surfaces, or even produce a video output, these can be added to the original base units.

Instruments are also available which are completely portable in that they are small, light and battery operated. An example is shown in Fig. 7.11.

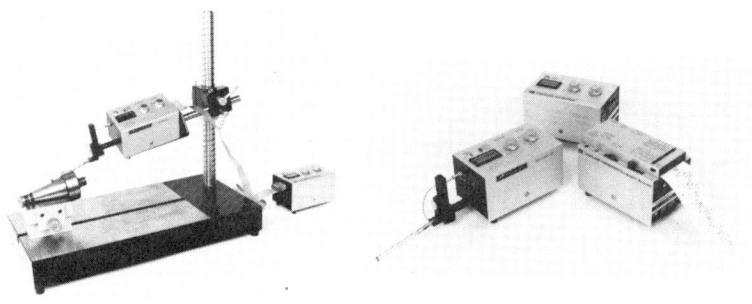

FIGURE 7.10

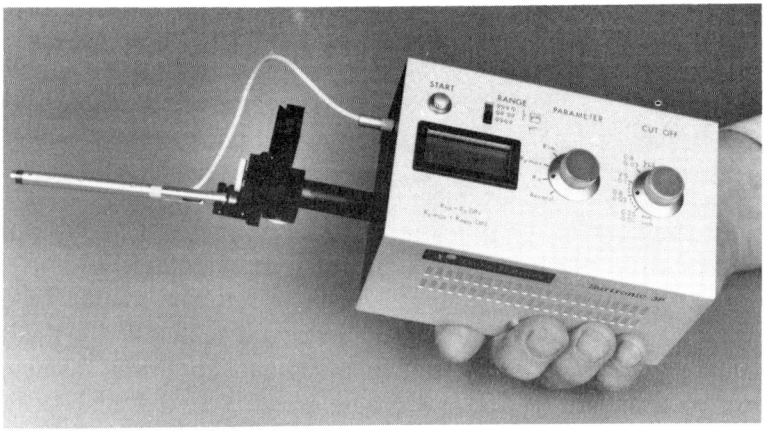

FIGURE 7.11

100 Flatness and Surface Finish

Surface finish instruments can also be obtained which use different principles. One such instrument is shown in Fig. 7.12.

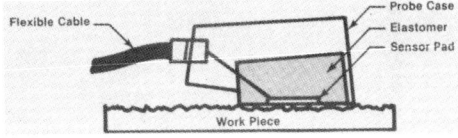

FIGURE 7.12

The measuring head of this instrument is static and is simply placed on the surface. The principle of operation is that the head forms a capacitor with the surface, the value of which depends on the air space between them, which, in turn, depends on the finish of the surface. Such an instrument is very easy to use, but it is a comparator and requires setting against a calibrated master of comparable finish.

7.5 USE OF REPLICAS

In spite of the versatility of surface finish measuring equipment, there will still be the occasional need to check the finish in positions which remain inaccessible. In such circumstances, the technique of obtaining a replica is useful.

This usually involves either pressing a heat softened plastic against the surface, or pouring a quick setting liquid into a mould formed by the surface to be checked as the base, and a room temperature mouldable material such as plasticine for the walls. A variety of plastic and resin compounds are available for this and, although the replica gives an inverted impression of the surface, the result obtained from a check on the replica will, for all practical purposes, be equivalent to that which would be obtained on the part itself.

7.6 FLATNESS

In addition to the elements which might be described as the roughness of the surface, it is often useful to be able to quantify the features which have a longer wave length. In fact, the elements which in Fig. 7.3 were identified as form. The main difference in measuring these features and those regarded as texture is that it is not practicable to use a stylus supported by a skid or shoe. The alternative is to move the stylus against a datum surface within the machine itself. Some surface finish machines have this option, but there are also machines which are designed specifically for this purpose, as the example illustrated in Fig. 7.13.

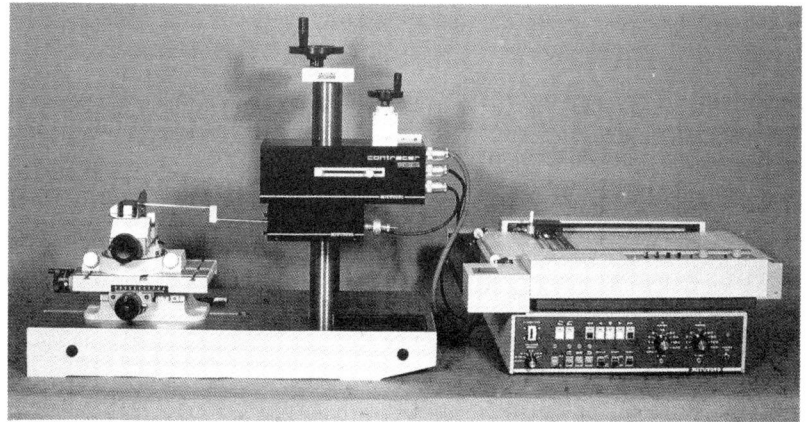

FIGURE 7.13

It is difficult to devise a parameter which will sensibly define flatness in a single figure, and the most useful output from these machines is a graphical recording which can then be compared with the specified tolerances. These are likely to be defined in geometrical form, as described in Chapter 3.

Chapter 8

Gear Teeth

In most engineering mechanisms, parts are to be found which contain features defined by parameters other than straightforward lengths and angles. Fortunately, the most difficult are quite common, and special equipment is readily available to deal with them.

There are few engineering products which do not have gears somewhere within them and the difficult metrology problem on such parts is, of course, measurement of the flanks of the gear teeth. This chapter is devoted to this particular problem.

8.1 TYPES OF GEARS

Gears may be designed in many forms, but the most common basic types are spur, helical, bevel and worm gears.

Spur gears have their teeth parallel to the axis of the gear in all planes.

Helical gears have teeth parallel to the axis when viewed along a tangent to the pitch circle, but at an angle when viewed in the direction of a radius. This form of gearing ensures that at least one pair of teeth is always in contact, and this gives quieter operation.

Bevel gears have teeth at an angle to the axis when viewed along a pitch circle tangent, but parallel when viewed along a radius. This form of gearing is used to transmit motion between two shafts at an angle. Bevel gears may have teeth which are also at an angle when viewed in a radial direction, and these are known as spiral bevel gears.

Worm gears consist of one spiral tooth, rather like a large thread, and are used for large speed reductions between two shafts at right angles.

The Involute Form 103

Gears for special purposes can be of exceptional shape with individually designed teeth, and gears can also combine to above basic types. Examples of various types of gears are shown in Fig. 8.1.

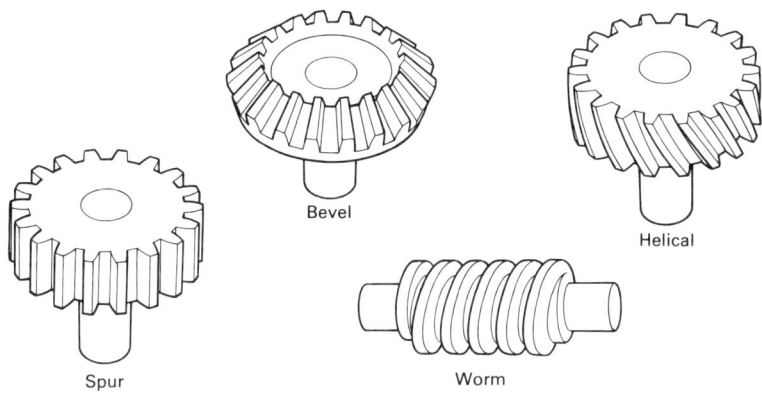

FIGURE 8.1

8.2 THE INVOLUTE FORM

The objective in designing gears is that, in operation, the contact between the working surfaces of mating gear teeth is rolling and not sliding. It is not possible to achieve this completely in practice, but in theory teeth of involute form satisfy the requirement, and this is the basis of tooth profile design.

Involute is the name given to the curve generated by a point on a straight line rolling without slip on a circle, as depicted in Fig. 8.2.

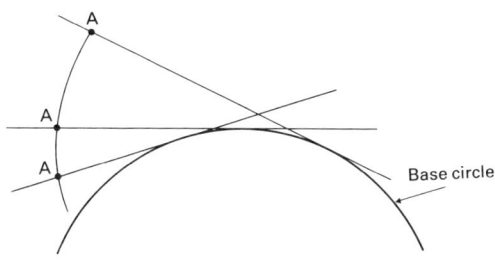

FIGURE 8.2

Gear Teeth

It can perhaps be more easily imagined as the curve traced by a point on a piece of string being unwound from a cylinder. This, in fact, is a practical method of drawing an involute shape.

The circle against which the straight line rolls is known as the "base" circle, and this parameter alone defines the shape of a particular involute.

Gears will normally be designed in pairs around two circles of such sizes that the sum of their radii is equal to the mounting distance between centres of the gears, and the ratio between the radii is the required ratio of speed between the gears when in operation. These circles are known as "pitch circles".

The other parameter needed to define a pair of gears is the "pressure angle". This is the angle between the tangent to the two base circles and the common tangent to the pitch circles. The relationship between the pitch circle, base circle and pressure angle is shown in Fig. 8.3.

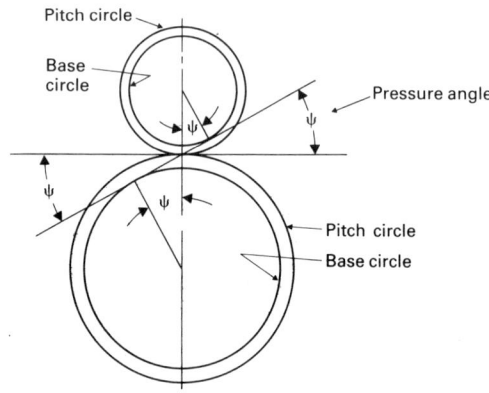

FIGURE 8.3

8.3 INVOLUTE MEASUREMENT

The basis of tooth profile measurement is to compare the tooth form with the theoretical involute profile. It is, however, possible to measure the form from first principles using a vertical rotating table and a height measuring instrument with a knife-edged stylus. The gear to be checked is mounted on the rotating table and the height gauge positioned so that the stylus contacts the tooth flank, as shown in Fig. 8.4.

Involute Measurement 105

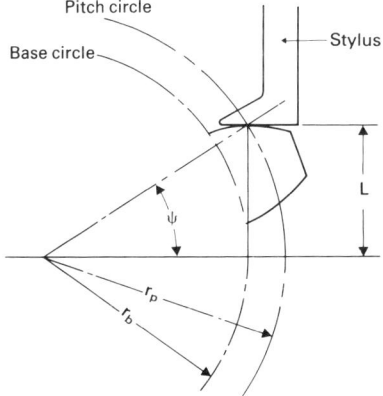

FIGURE 8.4

The profile is measured by recording the vertical movement of the height gauge stylus against angular rotation of the gear. If the involute is true, the vertical movement of the stylus will be equal to the arc of the base circle subtended by the angle of rotation (see Fig. 8.5).

This method is tedious and obviously unsuitable for checking gears, and even teeth, in quantity.

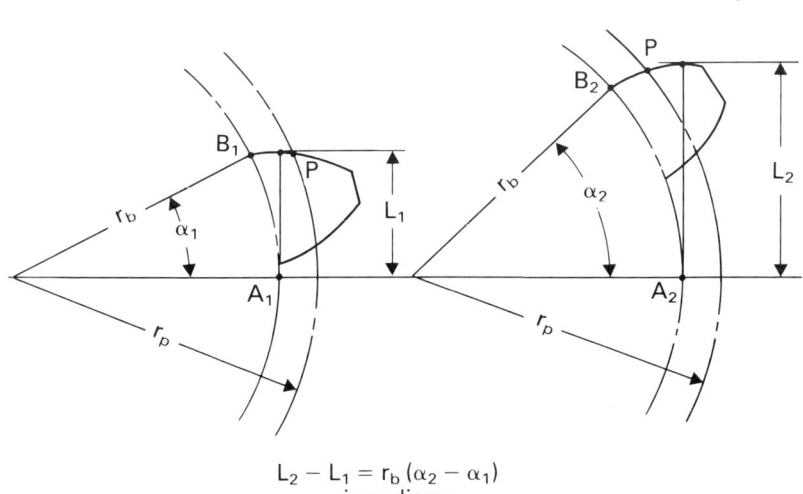

$$L_2 - L_1 = r_b (\alpha_2 - \alpha_1)$$
α in radians

FIGURE 8.5

106 Gear Teeth

Most of the machines available for checking tooth forms compare the tooth under test with an involute generated by the machine itself each time a check is made. Although a machine is described in Chapter 12 which does this using a computer, the majority of machines at present available generate the involute mechanically. The definition of an involute makes this practicable with reasonable ease, and the method is a particularly elegant solution to the problem of tooth form measurement. The means of generating the master involute is to roll the gear on its base circle against a flat plane. A point on the involute in this plane will remain stationary if the involute is correct. The principle is shown diagrammatically in Fig. 8.6.

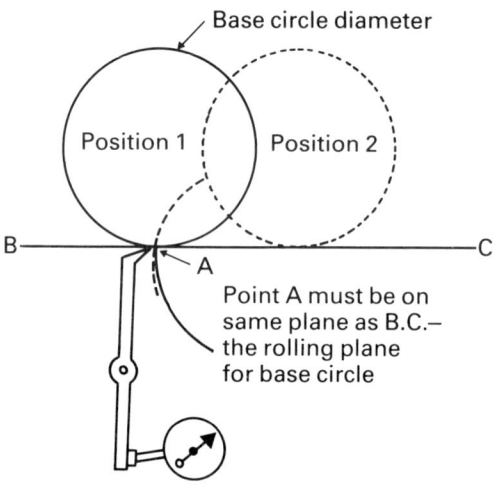

FIGURE 8.6

As the base circle rolls from position 1 to position 2, the locus of a point on a true involute will be stationary as it passes through the tangent to the base circle. Thus, if a stylus linked to an indicator is arranged to contact the tooth at this position, its movement will indicate deviations from a true involute.

A machine using this principle is shown in Fig. 8.7.

It consists essentially of a stylus linked to a straight edge which is loaded to keep in contact with the flank of the tooth being measured. The gear under test is rigidly mounted on the axis of a disc of the appropriate base circle diameter. For mechanical convenience, the axis of the gear and base circle disc is fixed, and

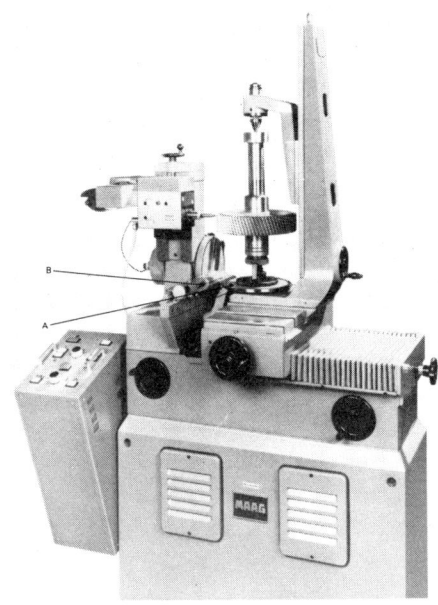

FIGURE 8.7

the base circle disc is rotated by movement of the straight edge. The base circle disc and the straight edge against which it runs can be seen at A and B respectively in Fig. 8.7.

In their simplest form, these machines require a different disc for each base circle diameter and this can be inconvenient unless large numbers of gears with the same tooth form are to be checked. The problem can be overcome, however, by means of the geometrical refinement shown in Fig. 8.8.

In this machine, the slide 1 on which the stylus 2 is mounted does not directly contact the base circle disc 3. It is connected to the straight edge 4 against which the base circle runs, by a pivoted link bar 5. The pivot 6 is at a distance from the straight edge 4 equal to the radius of the base circle disc 3. The distance between the slide 1 and straight edge 4 is variable, and this permits adjustment of the machine to accomodate any base circle diameter within its range from only one base disc.

Although it is possible to observe the movements of the stylus on an indicator, it is extremely difficult to judge exactly what these mean in terms of variations to the involute form. Consequently it is normal for the results of involute checking machines to be presented in graphical form. Again, the definition of the involute makes this

108 Gear Teeth

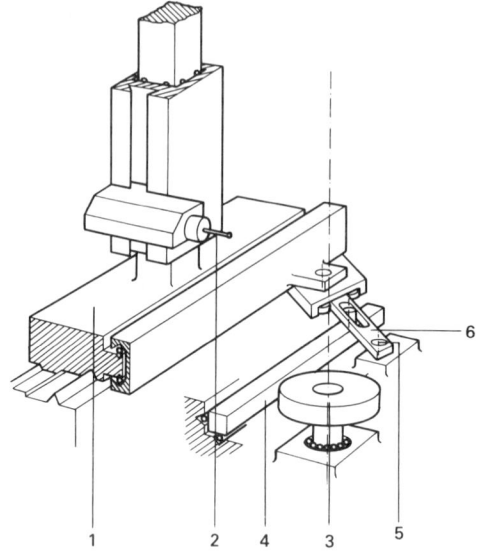

FIGURE 8.8

comparatively simple. In the machine shown in Fig. 8.9, the stylus is mechanically linked to a pen which records on a paper roll feed, again mechanically, by the movement of the straight edge against the base circle.

FIGURE 8.9

The recording obtained will be similar to the example in Fig. 8.10, the deviations from a straight line representing deviations from a true involute.

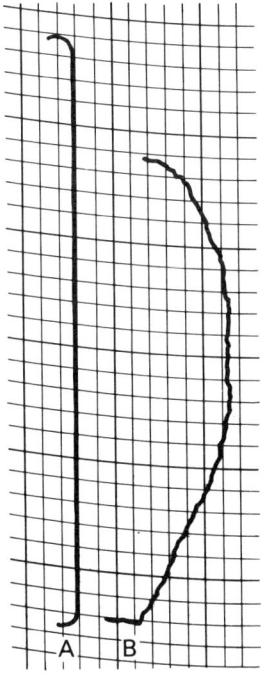

A = Diagram of a theoretical involute
B = Profile diagram of a tooth with tip and root relief

FIGURE 8.10

This recording, incidentally, is another example which has a magnification in one axis much greater than the other.

The trace in Fig. 8.10 shows a feature which is very common in the design of gear teeth in that material is missing at the tip and root. This is known as tip and root relief and it prevents fouling which might otherwise occur as teeth move into and out of mesh during operation.

The more elaborate machines, as the example in Fig. 8.7, also include provision for measuring the helix angle on helical gears. This is done by linking the rotation of the base circle against the straight edge with movement of the stylus parallel to the axis of the gear, as shown in Fig. 8.11.

The machine can be set so that a correct helix angle will produce a straight line graphical recording.

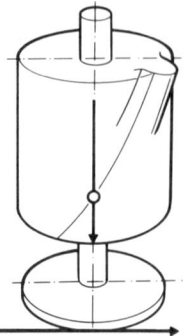

Figure 8.11

8.4 TOOTH THICKNESS AND PITCH

Both of these parameters are necessary to define gear teeth completely, and special purpose instruments are available to measure them.

Tooth thickness is likely to be specified as the thickness at a specific distance from the centre of the gear, most probably near the pitch circle diameter, at which point the tooth thickness is a maximum.

The instrument for measuring this is the quite simple tooth caliper shown in Fig. 8.12.

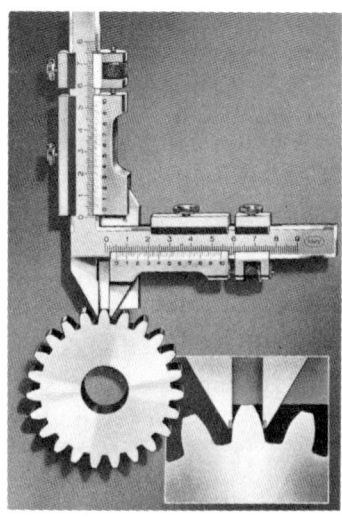

Figure 8.12

Tooth Thickness and Pitch 111

This might be described as a miniature vernier caliper with an adjustable stop to determine the points at which the anvils contact the gear flanks. In use, this stop is set by deducting the radius at which the thickness is to be measured from the outside radius of the gear, as in Fig. 8.13.

This method has the disadvantage that the measurement is taken on the corners of the caliper anvils, and errors can be introduced if these become worn. To avoid this, the anvil tips can incorporate hardened inserts, but allowance can be made for errors due to wear by using the caliper as a comparator. Figure 8.14 shows its calibration against a cylinder of such a diameter that it is approximately the width of the tooth at positions defined by the pressure angle.

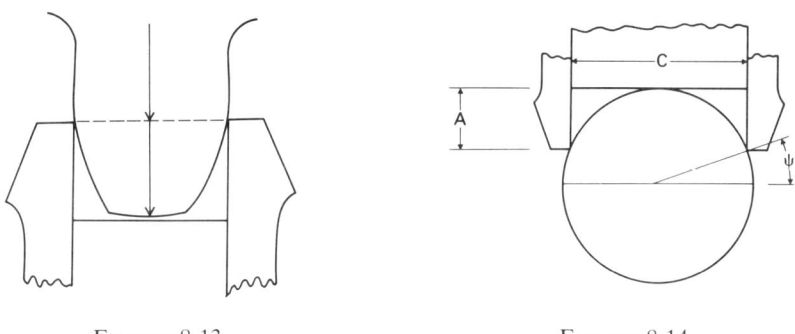

FIGURE 8.13   FIGURE 8.14

If ψ is the pressure angle and D the roller diameter, A must be set to a value $\frac{D}{2}(1-\sin\psi)$ and the value then obtained for C compared with the theoretical value $D\cos\psi$. The difference would be the correction applied when measuring a gear tooth.

Another method of measuring tooth thickness uses a normal vernier caliper over a number of teeth, as in Fig. 8.15.

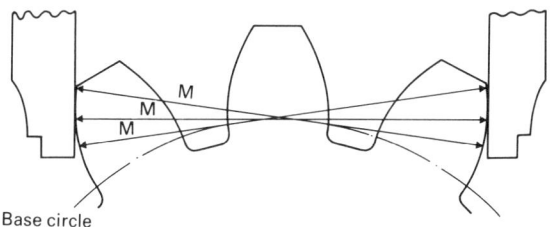

FIGURE 8.15

112 Gear Teeth

The number of teeth must be calculated so that the anvils of the vernier are tangential to the flanks of both the teeth they contact. The theoretically correct dimension will then be the length of the base circle arc between the origins of the two involute forms being contacted. This can be calculated from the rather complex formula:

$$M = 2r_p \cos\psi \left(\frac{T}{2r_p} + \tan\psi - \psi + \frac{\pi S}{N} \right)$$

where M is the theoretical vernier measurement,
r_p is the pitch circle radius,
ψ is the pressure angle (in radians),
T is the required tooth thickness at the pitch circle,
S is the number of spaces between the measured teeth,
N is the number of teeth on the gear.

A third method of obtaining a measure of tooth thickness is to measure the diameter of the gear over rollers placed in the teeth, as in Fig. 8.16. If the gear has an odd number of teeth, two rollers must be placed on one side and one on the other.

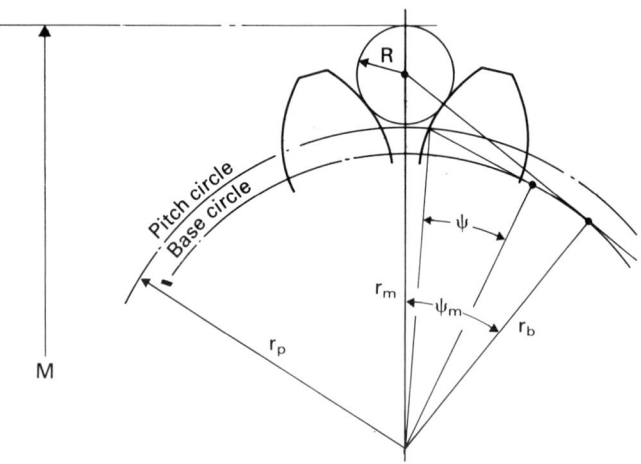

FIGURE 8.16

If the measured dimension over rollers is M and the radius of the rollers is R, the tooth thickness can then be obtained from the even more complex expression:

$$2r_p \left(\frac{\pi}{N} + \tan\psi_m - \psi_m - \frac{R}{r_p \cos\psi} - \tan\psi - \psi \right)$$

where $\psi_m = \cos^{-1}\dfrac{r_p \cos\psi}{r_m}$,

if the number of teeth is even, and $r_m = \dfrac{M-2R}{2}$

if the number of teeth is odd. $r_m = \dfrac{M-2R}{2\cos\dfrac{90°}{N}}$

The final parameter, the pitch of the teeth, is defined as the distance between corresponding points on adjacent teeth. This can be measured either absolutely or comparatively. To measure comparatively it is only necessary to mount the gear between centres and position a stop against one tooth, as shown diagramatically in Fig. 8.17 against tooth 1.

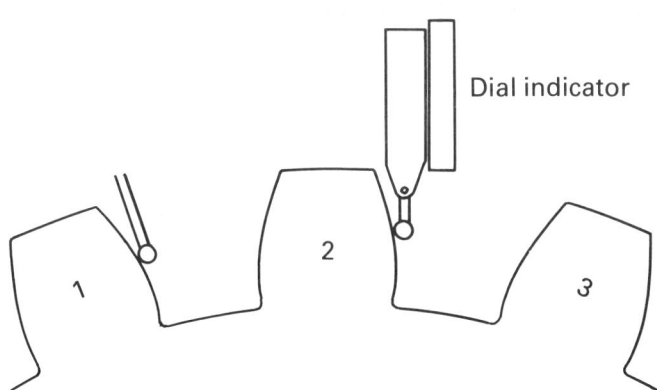

FIGURE 8.17

The stylus of a dial indicator is then positioned at approximately the same position on the adjacent tooth, 2, and the indicator set to zero. The gear is then moved so that tooth 2 is against the stop and the indicator registers the position on tooth 3. This process is repeated until all the teeth have been checked.

The obvious method of measuring the pitch absolutely is to mount the gear on an angular measuring device with a fixed indicator contacting one of the teeth, in a similar manner to the indicator against tooth 2 in Fig. 8.17. The indicator is set to zero and the gear then rotated through the angle theoretically required to bring the flank of the next tooth into the same position. The pitch error may then be read directly from the indicator.

Whilst this method is straightforward, very small errors in pitch can significantly affect the performance of a gear, and pitch measurement in this way requires very accurate angular measurement, generally better than 10 seconds of arc. For this reason, the method is not very practicable.

The two pitch measurements mentioned so far make use of general purpose equipment, but special purpose instruments are available, generally of the type shown in Fig. 8.18.

FIGURE 8.18

Essential features are an adjustable locating device rigidly attached to the body of the instrument and a stylus linked to an indicator. By setting the indicator to zero against one tooth and then obtaining a reading for all the others, comparative pitch measurements for the complete gear can be obtained. The instrument can measure absolute pitch if it is set to zero against a setting gauge, as shown in Fig. 8.19.

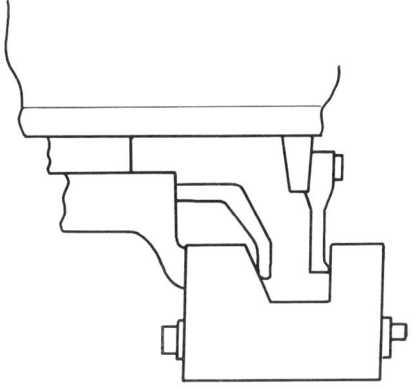

FIGURE 8.19

8.5 ROLLING TESTS

The gear checks so far described are tedious and time-consuming but necessary to obtain a full appreciation of the quality of a gear. However, the consistency of gear manufacturing methods is such that detailed checks of this sort need only be done on a sample basis.

If the acceptability of one tooth is established, then comparison with the rest of the teeth on a particular gear, and with the rest of the gears in a batch can be made quite rapidly by completely different means.

The simplest and most popular machine for checking gears in quantity is the rolling tester, as shown in Fig. 8.20.

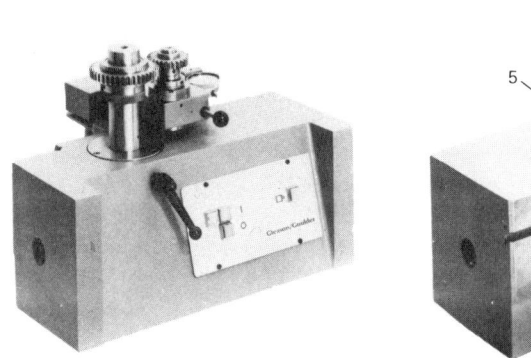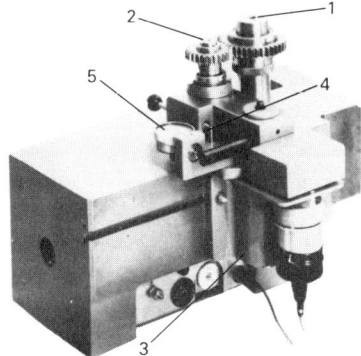

FIGURE 8.20

116   Gear Teeth

This consists essentially of a means of running together a pair of gears. These may be one gear under test and a master gear, or a pair of gears under test. In the machine in Fig. 8.20 one of the gears, normally the master, is mounted on a fixed spindle 1, whilst the other gear is mounted on spindle 2. This second spindle is mounted on a link pivoted at 3 and this permits the centre distance between the spindles to be varied. The link is loaded by spring 4 to keep the gears in mesh and, as the gears are rotated, the movement at spindle 2 is registered on a dial indicator 5.

Whilst the results can be taken from a dial indicator, a graphical recording is a much more valuable output. A more sophisticated machine, which gives a graphical recording of centre distance variations as the gears rotate, is shown in Fig. 8.21.

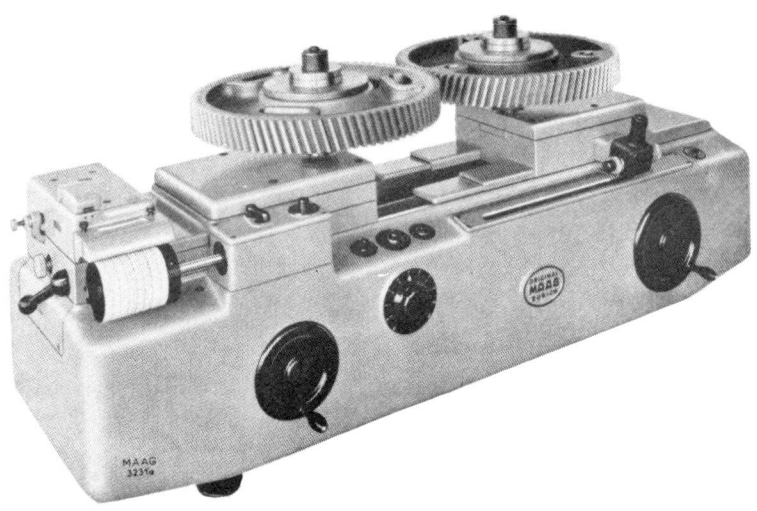

FIGURE 8.21

Even with a graphical recording, the results obtained from such a check are not easy to interpret because the movement of the centres of the gears relative to each other is influenced by so many factors. The method is, however, a valuable accept or reject check because it is not difficult to specify an acceptable trace. If the trace is not acceptable, experience may indicate the likely cause from the shape of the trace. Figure 8.22 shows examples of traces likely to be obtained for specific reasons.

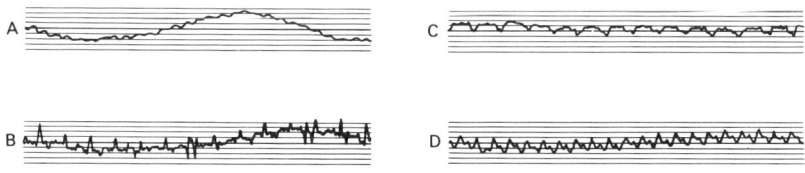

FIGURE 8.22

Trace A is an indication of concentricity errors in one or both gears, whilst trace B would be an indication of pitch errors between adjacent teeth. Trace C shows the shape likely to result from tooth profile errors and D might be expected if errors in pressure angle exist.

If the trace does not disclose any obvious cause of a problem it will be necessary to resort to other checks to analyse the reason.

It will be appreciated that unless a gear train is intended to operate in both directions, only one flank of the gear tooth performs a useful function. Designs normally allow a fairly large clearance, perhaps 0.5 mm, between non-working flanks when the gears are mounted at working distance. This clearance is defined as the Pitch Line Clearance, but is generally known colloquially as the "backlash".

Rolling testers of the type described operate with the gears loaded in mesh and the test is carried out without backlash. Such machines are comparatively inexpensive, but have the disadvantage that the readings obtained are influenced by the shape of both working and non-working flanks. This not only confuses the result of the test but could also cause the rejection of a gear for deficiencies on the non-working flanks which, in practice, are of no consequence.

This problem is overcome by a machine of the type shown in Fig. 8.23 which carries out a rolling test with only the working flanks of the gears in contact.

The principle is different in that the gears are mounted at their theoretically correct centre distance. The test is then carried out by rotating the gears against the load in the working direction, and the machine records variations in the rotation of one gear relative to the other.

The principle of the machine is illustrated in Fig. 8.24.

Each of the gear mountings is rigidly attached to an optical grating. These are electronically linked and this link includes multiplying and dividing circuits which permit gear trains of any ratio to be tested.

118 Gear Teeth

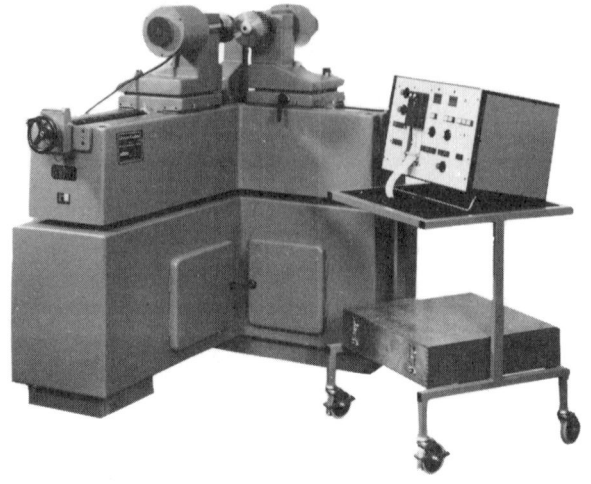

FIGURE 8.23

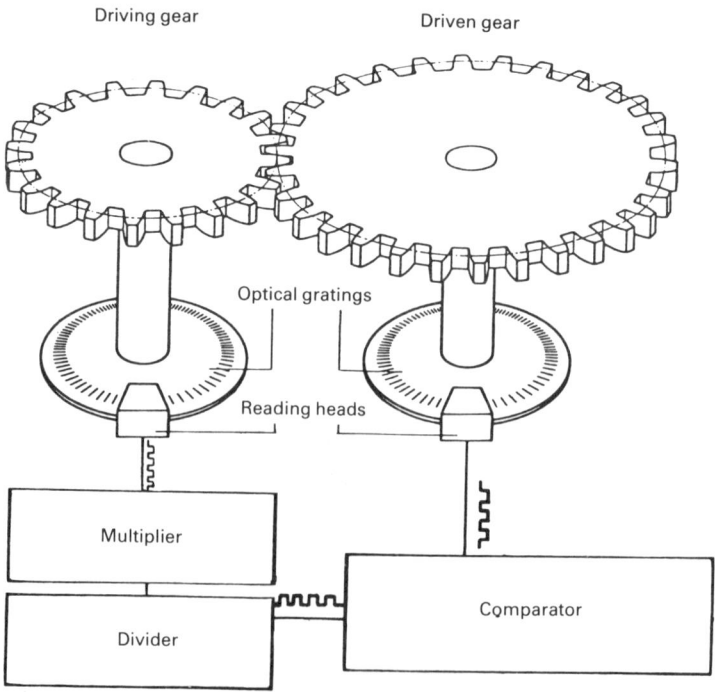

FIGURE 8.24

The results again appear as a graphical recording and although the traces resemble those obtained from a dual flank machine, it is normally easier to identify specific causes of errors.

Although the rolling tests described have implied the testing of simple spur gears, the same principles apply to gears of most types. Both dual and single flank machines can be obtained with provision for varying the angle between the spindles so that bevel gears can be tested.

8.6 TOOTH MARKING TESTS

The single flank tester may be used in a different way by applying marking ink to the teeth of one of the gears and observing the transferred marking on the other. For successful results this form of testing requires a power drive and a brake, so that a reasonable load can be applied to the gears, ideally approaching the load under working conditions.

Since measurement of tooth profiles of bevel gears is particularly difficult, this form of test is a useful alternative for this type of gear. The main features of a machine designed for this purpose are shown in Fig. 8.25.

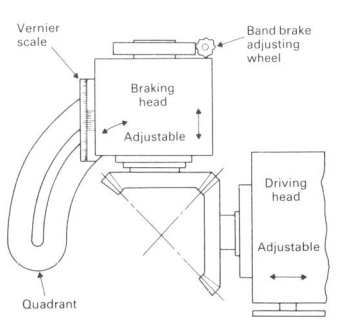
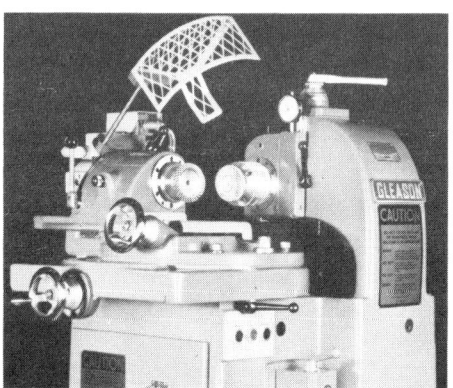

FIGURE 8.25

Examples of tooth markings likely to be obtained are shown in Fig. 8.26, and whilst these might be identified absolutely to specific errors, in practice the desired marking is usually obtained by trial and error.

120 Gear Teeth

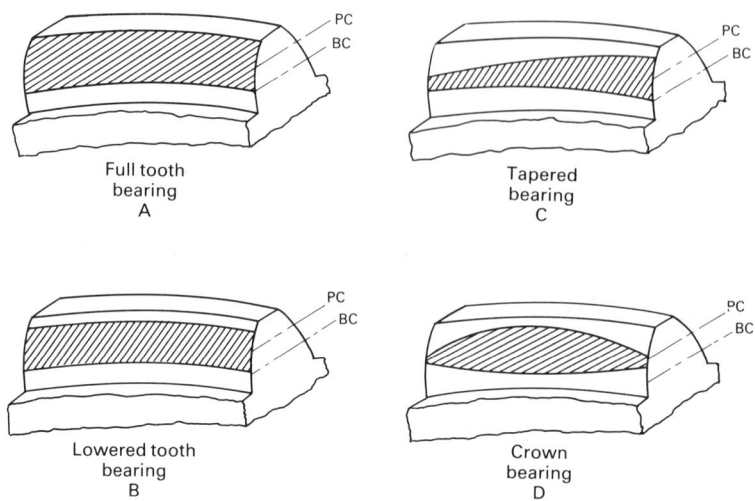

FIGURE 8.26

This is because, under working loads, the mounting positions of the gears are likely to change due to casing deflections and irregular taking up of clearance. The use of trial and error methods to define markings is in fact a means of building in compensations for operational deviations.

Chapter 9

Measurement of Contoured Surfaces

The last chapter was devoted to the consideration of an example of a widely used surface in engineering which is not nominally either flat, cylindrical or spherical. In spite of this, the measurement of gear teeth is simplified in many respects because it is another regular geometrical form.

Engineering products, however, often contain surfaces which are generated to perform a specific function, or to be aesthetically pleasing, and it is unlikely that these will be capable of definition by any reasonably simple algebraic formula.

The inspection of such features has, in principle, the same objectives as any other inspection, i.e. to check the degree of conformance to the specified requirement. It is consequently likely that the way in which the feature is specified will give an indication of the way in which it might be inspected.

9.1 OPTICAL PROJECTION

If the profile is only two dimensional, optical projection, as described in Chapter 11 could be the obvious and most economical method. In order to use this method for a contoured surface, it is necessary to produce a master layout of appropriate magnification. It is possible that such a layout will be the means of specifying the shape, in fact this method is to be encouraged if it is apparent at the design stage that projection will be a suitable method of inspection. Recent developments in plastic materials have resulted in the availability of transparent sheet which is sufficiently rigid and stable to be used with confidence for specifying profiles with tolerances as close as 0.1 mm.

The accuracy required is, of course, a major consideration in determining the validity of projection as an inspection method.

122 Measurement of Contoured Surfaces

Under the most favourable conditions, an adequately sharp image at a magnification of 50× may be obtained, and this can be measured to an accuracy of about 0.5 mm on the screen, giving an accuracy of measurement on the workpiece of 0.01 mm.

In addition to the simple shadow and surface projection techniques, special purpose machines using developments of these techniques are available for checking contoured surfaces which are not so readily accessible or do not have an exposed cross-section.

The illustration in Fig. 9.1 indicates an ingenious means of examining a cross-section which is not necessarily exposed as an end face.

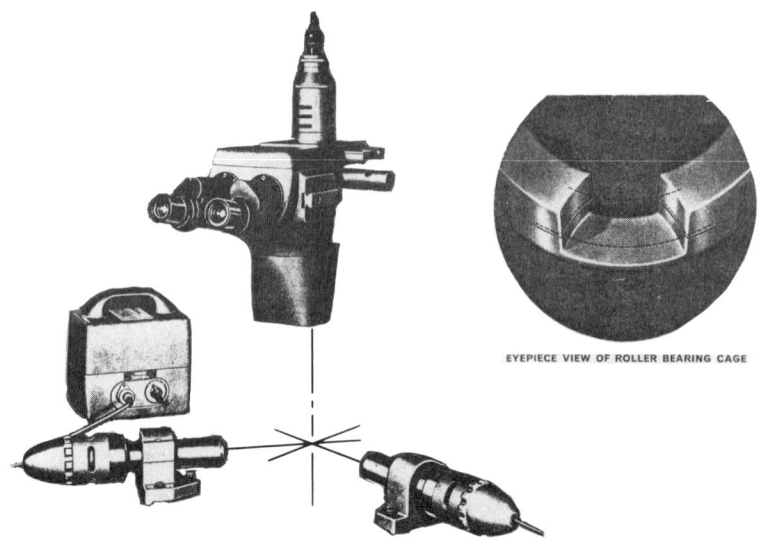

FIGURE 9.1

This instrument focuses a line of light on the cross-section to be measured. The part is then viewed through a microscope in a direction perpendicular to the section. The image in the eyepiece will then be effectively a cross-section at the line of light. An important requirement is that the objective lens of the system is sufficiently large to accept light which might otherwise be obscured by any projections on the part between the section and the lens. The diagram in Fig. 9.2 shows that, whilst the light directed at the section will be randomly reflected in all directions, only the proportion striking the lens contributes to the visible image.

The same principle can be used for projection on a screen, but in this case, since much more light is required to produce an adequate screen image, a large diameter lens is even more important. The

Optical Projection 123

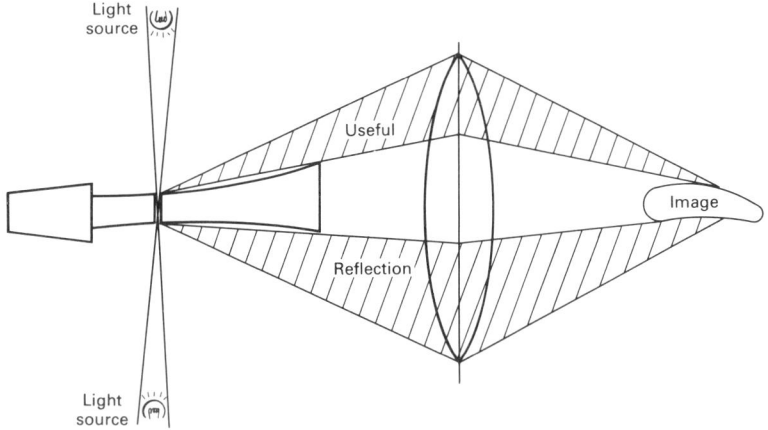

Figure 9.2

image is, incidentally, likely to be improved if the surface of the part is treated with a reflecting material such as white paint or a dusting of white chalk.

This projection technique is particularly suitable for the aerofoil forms of turbine and compressor blades, and a projector specifically designed for this purpose is shown in Fig. 9.3.

Figure 9.3

124 Measurement of Contoured Surfaces

It may prove beneficial to project on to the section a broad band of light instead of a thin line. If this technique is used, the section required must be at one edge of the band, and that section must be sharply in focus. The image obtained will then have a diffused region inside or outside the section line, depending on which edge of the light band is used to define it.

The image obtained by projecting the cross-section of a fir tree root form in this way is shown in Fig. 9.4.

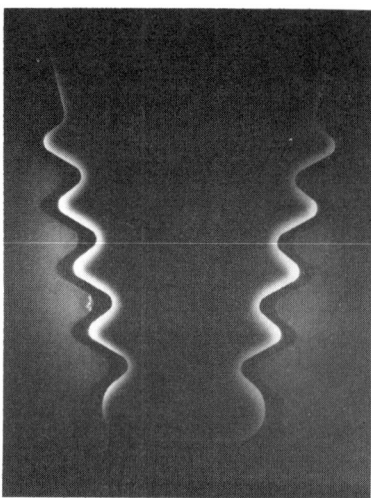

FIGURE 9.4

Normal shadow projectors may also be used in conjunction with mechanical devices to produce cross-section enlargements. Perhaps the simplest device for doing this is of the type shown in Fig. 9.5.

This consists of a vertical and a horizontal stylus mounted above each other on a vertical stand. The item to be measured is mounted on a surface table so that the section concerned is horizontal. The horizontal stylus is then positioned at the same height as this section and it traces around the section by moving the stand on the surface table by hand. At the same time, the vertical stylus is maintained by spring loading in contact with a horizontally mounted glass slide coated with opaque lacquer. As the horizontal stylus follows the section being measured, the vertical stylus cuts a corresponding image on the glass slide, which may then be projected in the normal way.

It is essential that the two styli are precisely positioned one above the other and this can be checked by tracing a rectangular or circular object of known size.

Optical Projection 125

Figure 9.5

This method is a particularly cheap and simple way of checking irregular shapes in small quantities.

A development of this technique is used in the projector shown in Fig. 9.6.

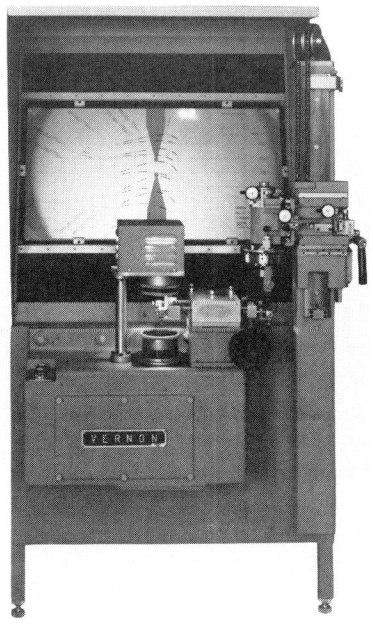

Figure 9.6

126 Measurement of Contoured Surfaces

In this machine, a spherical stylus which traverses the section is built in. It is rigidly linked to an exactly similar stylus, or to a graticule inscribed with a circle of the same diameter. As the stylus moves around the section, the image of the second stylus, or the graticule, is projected on a screen, and its movement checked against an enlarged master layout. This can be seen on the screen shown in Fig. 9.6, which, incidentally, is of a projector having two independent styli so that both surfaces of an aerofoil form can be checked at the same time.

The projector shown in Fig. 9.7 adopts a slightly different principle, in that the floating carriage to which the stylus is linked holds the projection lens.

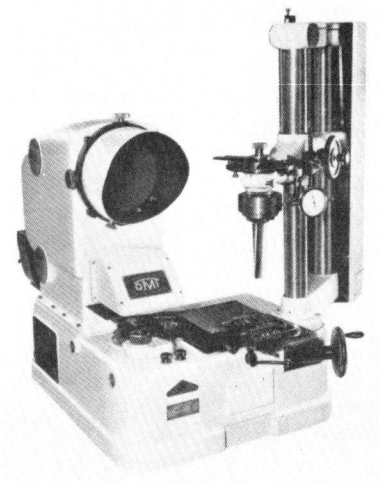

FIGURE 9.7

A fixed graticule of the desired shape at normal size is projected through this moving lens and this causes the projected image to move across the screen as the stylus moves around the section. The screen in this case contains an enlargement of the stylus head, possibly surrounded by a similarly magnified tolerance zone. If the shape is correct, the desired shape line on the graticule will appear to move between the stylus and tolerance zone lines on the screen, as indicated in Fig. 9.8.

Although there is a problem of preparing a normal size graticule, this method makes it possible to use a large magnification, perhaps 30×, on a large profile using only a small screen.

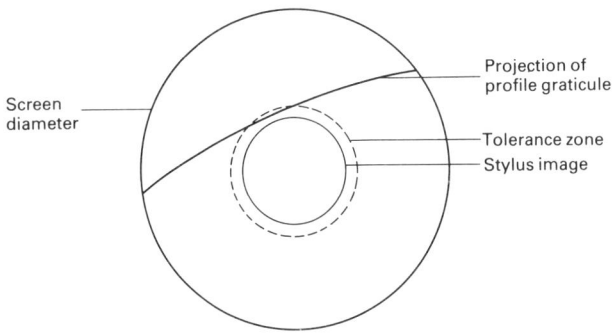

FIGURE 9.8

9.2 TEMPLATES

A simple, but not very accurate, means of checking shapes is by using templates.

The radius gauges shown in Fig. 9.9 are a popular use of this method, which involves merely placing the gauge against the surface and observing the extent to which it is in contact.

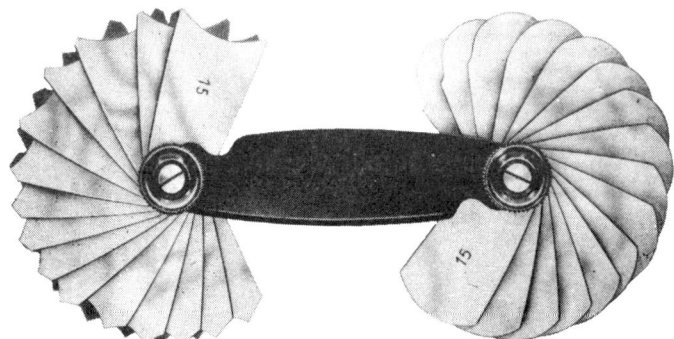

FIGURE 9.9

The method is quantitative only by estimation and the accuracy of the result depends largely on the skill of the user, but the method can be enhanced by mounting the part and the templates in a fixture which ensures their correct positioning relative to each other. In this form, templates are used for checking quite complex shapes.

A development of the template technique is seen in the device in Fig. 9.10.

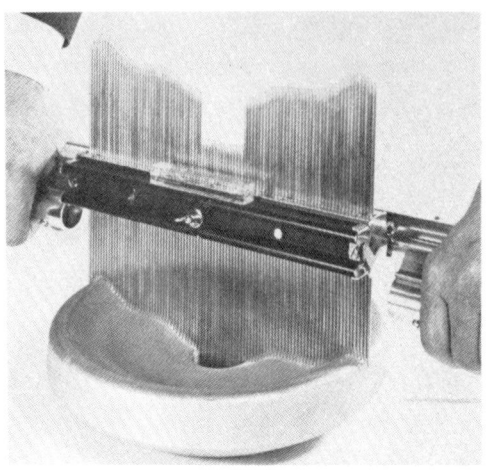

FIGURE 9.10

This actually creates a template at one end of a large number of equal length needles by pressing their other ends into contact with the surface. This "template" can then be compared with the desired shape, either directly or by projection.

9.3 MULTI-PROBE MEASUREMENT

An alternative to measuring the complete profile of a contoured surface is to carry out a series of checks at specific points. In order to do this the position of the nominated points must either be the means of specifying the contour, which is quite common, or they must be calculated.

The positions can be specified in two or three dimensions, and with this method, unlike projection, three dimensions can be handled with little more difficulty than two.

Measurements can be made in this way on the multi-axis measuring machines described in Chapter 5. This involves using one vertical measuring stylus to take readings at various positions in the horizontal plane, as shown in Fig. 9.11.

The alternative is to use a number of vertical indicators positioned in a fixture so that they contact the surface at the desired positions, all at the same time. In theory, the vertical indicators could be dial indicators as in the multi-dimension inspection fixture mentioned in Chapter 5, but since all the dimensions are required to be on one surface of the part, this is likely to be impracticable because of

Multi-probe Measurement   129

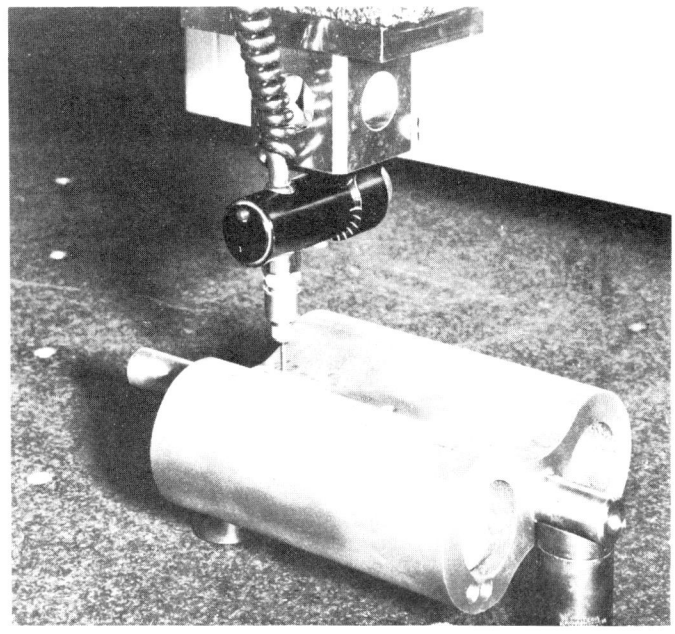

Figure 9.11

insufficient space. In practice, such a machine is made possible by using remote reading pneumatic or electronic indicators. A machine of this type is shown in Fig. 9.12.

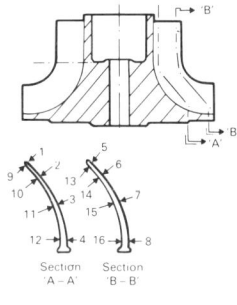

Figure 9.12

130 Measurement of Contoured Surfaces

The results may again be presented in a variety of forms. For quantity production the most useful might be a coloured light display, as in Fig. 9.13, in which a green light indicates a fully acceptable dimension, a yellow light shows incorrect but in a direction caused by excess material, and a red light incorrect due to insufficient material. The significance of the red and yellow is, of course, that the yellow condition is correctable whilst the red is not.

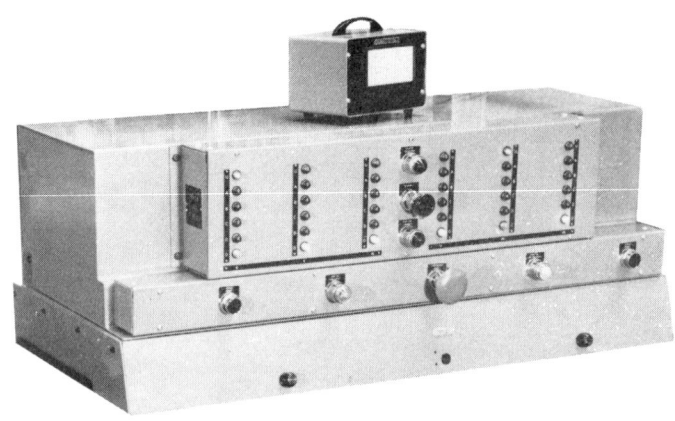

FIGURE 9.13

Such a display may also have one overall light indicating the total condition of the surface.

Alternatively, or in addition, means may be provided for displaying the actual deviation from the required size. This may be on separate analogue indicators for each dimension, one analogue or digital indicator switched between positions, or analogue column indicators as shown in Fig. 9.14.

Indicators of this type may be of liquid columns or electronic simulations.

If electrical indicators are used, the output can be processed to produce output in endless variety. A spectacular example is the three-dimensional pictorial impression shown in Fig. 9.15.

This diagram indicates pictorially the vertical deviation from the nominal position of each of the points checked. It must again be realised, however, that the picture is distorted because of the large difference between the magnifications used in the three planes.

Multi-probe Measurement 131

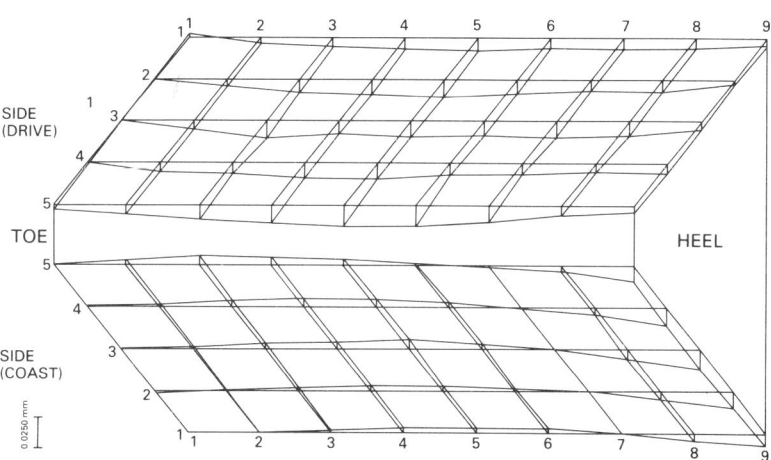

Figure 9.14

Figure 9.15

Chapter 10

Screw Threads

There are very few engineering products which are not held together by the ubiquitous nut and bolt. Even if the fastening is not exactly a nut and bolt, it is possibly a nut and stud or a set bolt. In any case the soundness of the product is likely to be dependent somewhere on a screw thread.

10.1 DEFINITION OF A SCREW THREAD

An enlarged screw thread is shown in Fig. 10.1 and the seven elements by which it is defined are identified. They are:

>   major diameter,
>   minor diameter,
>   effective diameter,
>   pitch,
>   flank angle,
>   form at crest,
>   form at root.

The major and minor diameters and crest and root forms are self-explanatory, but the others perhaps require a more detailed definition.

It must first of all be remembered that all the thread parameters are measured either parallel or normal to the axis of the cylinder on which the thread is produced. Since the thread itself is a helix, the results will not be the same if measurements are taken parallel or normal to the thread itself (see Fig. 10.2).

Flank Angle

As shown in Fig. 10.1, the total flank angle is the angle between two adjacent thread surfaces, but for performance assessment the

Definition of a Screw Thread

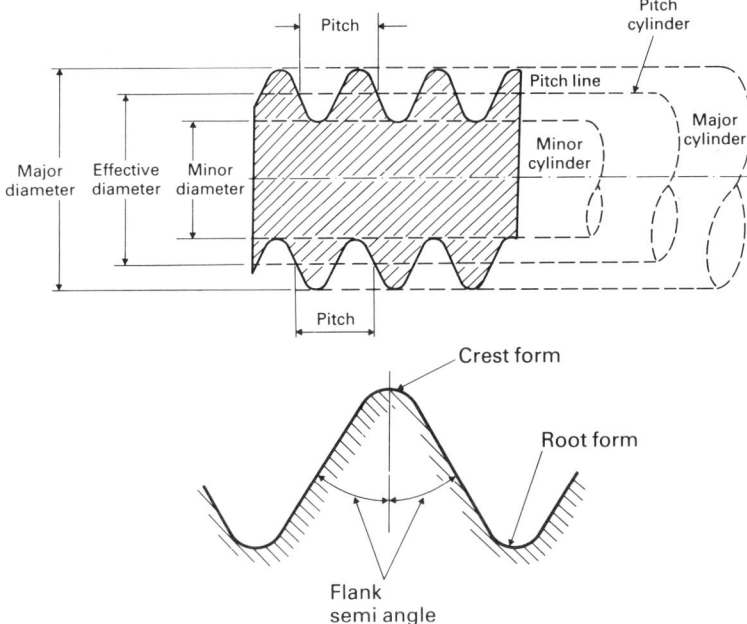

FIGURE 10.1

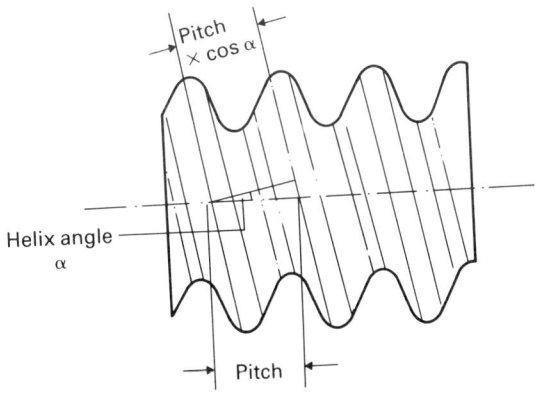

FIGURE 10.2

angle of each flank of the thread is important and it is usual to measure the angle as two semi-angles, measured from the flank concerned to a perpendicular to the axis.

Effective Diameter

The effective diameter and the pitch are related, and the effective diameter can be defined as the diameter of the cylinder which intersects the surfaces of adjacent threads, at a position such that the distance between the intersecting points is equal to half the pitch.

Pitch

The pitch is the distance measured parallel to the axis between corresponding points on adjacent threads. It would nominally be measured on the effective diameter, but the same reading will be obtained anywhere along the flat portion of the thread flank.

A thread can, of course, be an original design and each of the parameters selected to suit the particular purpose, and this might still be done in cases where a screw thread is required to be an integral part of a component. Normally, however, threads are selected from standard ranges; in fact, threads are perhaps the earliest and most widely used standards.

10.2 STANDARD THREADS

The most traditional Imperial standard thread is the Whitworth, known as BSW. By precision engineering standards this is a coarse thread of a somewhat agricultural nature. To cater for the need for greater precision a finer thread known as BSF was introduced, and these two threads formed the basis of the Imperial thread standards. In order to cater for very small threads, a numbered series known as BA threads existed and the range is further confused by a series known as BSP which is used almost exclusively for pipe end fittings.

The BSW and BSF ranges are identified in size by the major diameter of the thread, the BA size by a number which in itself cannot be related to a particular dimension, and the BSP size by the bore diameter of the pipe on which the thread is used. A range of threads corresponding to BSW and BSF also existed in the United States, but their size was identified by the dimension across the flats of the nuts or bolt head. There were also corresponding ranges of metric threads.

In comparatively recent years an attempt has been made to achieve an acceptable international standard by introducing the unified series. This includes ranges known as UNC and UNF which correspond to the BSW and BSF series. They are defined in size by the major diameter and, even for metric designs, it is normal to use sizes defined in fractions of inches. Unified threads use the metric flank angle of 60° as opposed to the Imperial flank angle of 55°.

10.3 THREAD MEASUREMENT

Facilities and techniques for thread measurement cover the range between the extremes of simplicity and sophistication.

Normal practice would be to carry out a detailed check to ensure that the process for producing the thread was adequate and to repeat this detailed check at regular intervals. These checks would be quantitative, but production quantities in between would be checked by simple accept or reject methods.

For detailed checks of the highest precision a toolmaker's microscope, or similar instrument, as described in Chapter 13, might be used. Such checks would be required regularly on the gauges used for production checking. Such an instrument is capable of examining all the parameters of the thread form, and in order to check the flank angles a goniometric microscope can be used. This has two reticles which can be rotated with respect to each other. These reticles can be aligned to the axis and the flank of the thread and appear as in Fig. 10.3.

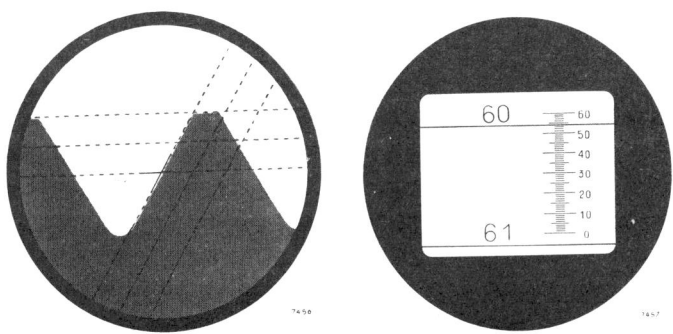

FIGURE 10.3

The angle between these two reticles can also be read in the microscope, as shown in Fig. 10.3, which shows a value of 60° 54'.

Whilst not having the same facilities for obtaining a quantified measurement of the angle, an alternative to the goniometric microscope is a normal full form reticle of the type shown in Fig. 13.13. The thread can also be checked for these features using an enlarged layout on similar lines in a normal horizontal or vertical projector.

For obtaining quantified measurements on other dimensions of the thread, a micrometer together with suitable accessories can often be successfully used, particularly if the thread is small. The major diameter can be measured with the micrometer as though the thread were a plain cylinder, but it must be remembered that any diameter measurements will be centralised on one thread on one side and span two threads on the other. The micrometer anvils, therefore, must be able to cover the thread pitch comfortably.

To measure the effective diameter, thread measuring cylinders are used. These are very simple in shape and are often supplied in sets to cover various thread forms, as shown in Fig. 10.4.

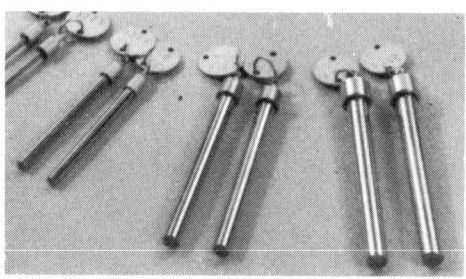

FIGURE 10.4

These are placed into the thread form and are of such a size that they will contact a straight portion of a flank at about its mid-point, as shown in Fig. 10.5.

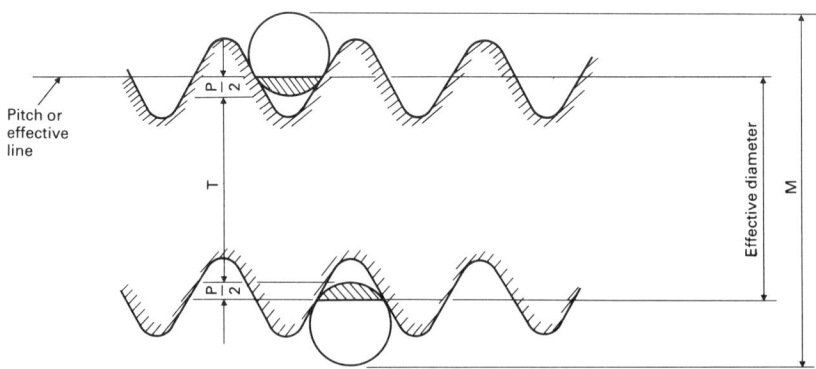

FIGURE 10.5

It will be seen that the cylinders sink into the thread groove beyond the effective diameter by an amount $P/2$. If the measurement over the cylinders is M and the diameter of the cylinders is D, then:

$$\text{effective diameter} = M - 2D + P.$$

Thread Measurement 137

If the measurement is being carried out with a hand-held micrometer, two cylinders are required on one side and one on the other. This can present an awkward problem in manipulating the cylinders, micrometer and thread at the same time. To facilitate this the cylinders can be obtained in sets, mounted in cages, as shown in Fig. 10.6.

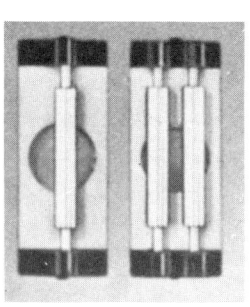
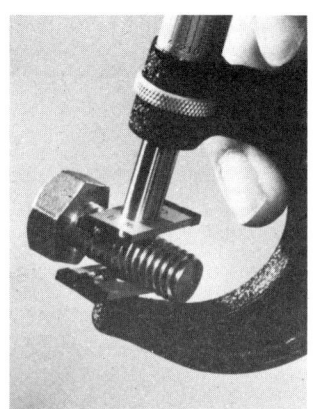

FIGURE 10.6

Whilst the ideal size for the cylinders is that which produces contact on the effective diameter, this is not particularly critical and the tabulation in Fig. 10.7 gives an indication of acceptable ranges.

Form of thread	Diameter of measuring wires		
	Maximum	'Best size'	Minimum
Whitworth – 55° B.A. – 47½° Metric (S.I.) – 60° Unified – 60°	0.853 × p 0.730 × p 1.010 × p 0.938 × p	0.564 × p 0.546 × p 0.577 × p 0.577 × p	0.506 × p 0.498 × p 0.505 × p 0.505 × p

FIGURE 10.7

The exact point, therefore, at which the cylinder contacts the thread is not particularly important, and provided the diameter of the thread is accurately known, the value P, and thus the effective diameter, can be calculated as shown in Fig. 10.8.

138 Screw Threads

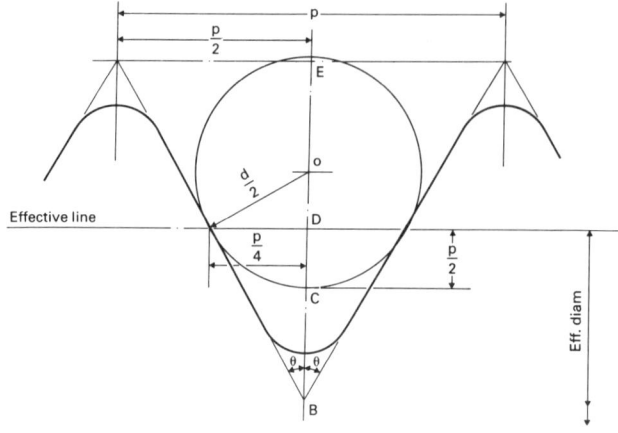

FIGURE 10.8

If d = diameter of cylinder,
p = pitch,
θ = flank semi-angle,

Then $P = \dfrac{p}{2}\cot\theta - d(\operatorname{cosec}\theta - 1)$.

It would not normally be necessary to perform this calculation, since the cylinders are generally supplied calibrated for use with a particular thread and are marked with the required P value, but it does make it possible to use cylinders for threads other than those for which they are calibrated.

Whilst thread measuring cylinders can be used with a normal micrometer, they can give more precise results if used with a special type of micrometer, sometimes known as a floating carriage micrometer, which is particularly useful for thread measurement.

An example is shown in Fig. 10.9, and it consists of a base on which are mounted two carriages at right angles, both free to move on ball tracks. The lower carriage has centres for mounting the thread to be measured, and the upper carriage has a micrometer with a large diameter thimble. Since the machine itself ensures that measurement is at right angles to the axis, effective diameters can be measured with only two cylinders. These are suspended from hooks provided for the purpose, as shown in the sketch in Fig. 10.10.

The machine is used as a comparator and is set against a cylinder of a size near to the effective diameter to be measured. A plain plug gauge can be used for this purpose.

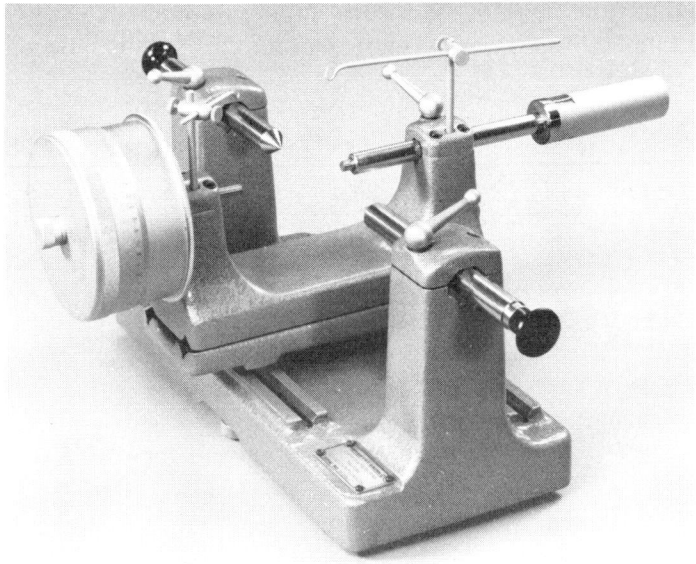

FIGURE 10.9

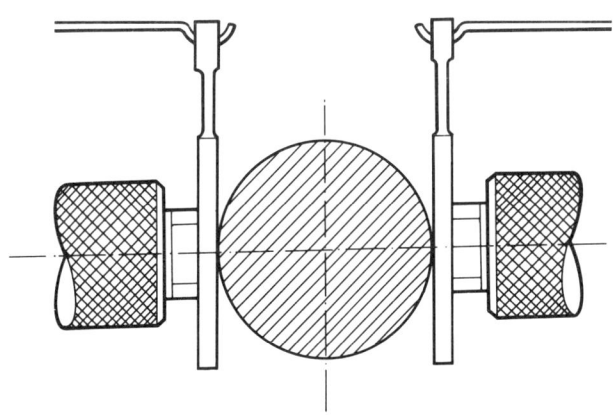

FIGURE 10.10

When a high order of accuracy is required, one source of error is the compression of the thread cylinders. This can be significant in effective diameter measurement because six point contacts between straight lines and circles are involved. When using the floating carriage machine, if the datum is established with the thread cylinders in place between the micrometer anvils and the gauge, the compression effect is largely compensated for.

140 Screw Threads

A similar technique is used for checking the minor diameter of a thread, but in this case, instead of cylinders, prisms as shown in Fig. 10.11 are used.

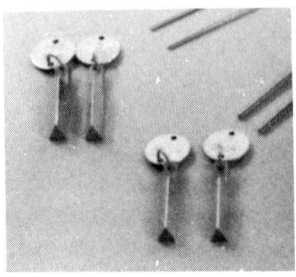

FIGURE 10.11

These have a sharp edge which can reach the smallest diameter without fouling the root radius.

For the measurement of pitch, it is necessary to use a measuring machine, such as a toolmaker's microscope, as described in Chapter 13, which can traverse along the axis of the thread and has provision for accurately recording movements along this axis.

A comprehensive pitch check would record the axial position of each thread along a line parallel to the axis, and to repeat the check with the thread turned through 180°. Each pitch can be located for this check either by a stylus with a ball end which contacts both flanks at about the effective diameter, or by a microscope with a reticle of the thread form being checked, as shown in Fig. 10.12.

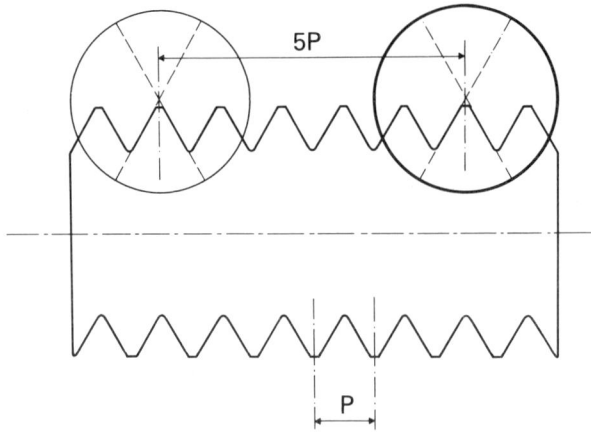

FIGURE 10.12

10.4 VIRTUAL EFFECTIVE DIAMETER

Assuming all the parameters of an internal and an external thread of the same form are correct, these threads would precisely fit together if they had the same effective diameter. Apart from some small clearance which would be necessary to cater for lubrication, etc., this condition will never be satisfied in practice because of errors in pitch and flank angle. Both these errors result in a fit of an external thread only being possible in an internal thread of larger effective diameter. This larger value, which includes compensation for errors in other parameters of the thread, is known as the virtual effective diameter.

10.5 INTERNAL THREADS

The measurement methods so far described are particularly appropriate for external, or male, threads. Internal threads present certain additional problems in respect of checking thread shape and of measuring the effective diameter.

Although an internal thread cannot be projected directly, it is possible to check flank angles and root and crest forms by projecting replicas. This technique is also mentioned in connection with surface finish checks and generally involves "casting" an inverted replica usually with a quick setting plastic material.

Measurement of the effective diameter requires more elaborate equipment, and an example is shown in Fig. 10.13.

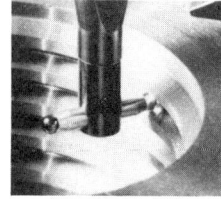

FIGURE 10.13

This uses a double spherical ended stylus in conjunction with a microscope, which locates its position along the axis of a measuring machine. Again, a machine of the toolmaker's microscope type is suitable. The microscope is used to locate the stylus at opposite ends of a diameter, and the effective diameter is obtained by adding the distance moved to a stylus constant.

10.6 MEASUREMENT OF PRODUCTION THREADS

Having established that the manufacturing process is producing threads with good crest and root shapes and flank angles, the principle of checks on production quantities is to ensure satisfactory mating of external and internal threads. This is achieved primarily by means of gauges which simulate the mating threaded component.

This is a particularly useful method for internal threads because of the difficulty of measuring them in other ways, and it involves very widely used plug gauges.

10.7 PLUG GAUGES

Figure 10.14 shows examples of plug gauges for checking internal threads. Plain plugs are provided for checking the minor, or core, diameter, whilst threaded gauges are used for other features. All use the "GO" or "NO GO" principle, which is simply that the thread is acceptable if the "GO" gauge will enter the thread and the "NO GO" gauge will not.

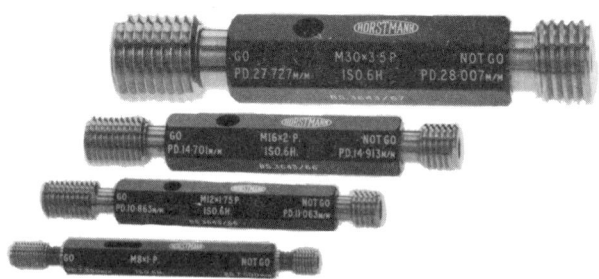

FIGURE 10.14

The use of the plain plugs to check the minor diameter is obvious, but the design of gauges to check other features is worth further consideration. The "GO" gauge has the form shown in Fig. 10.15a, and it will be noted that the outside diameter is truncated at the minimum permissible size for the major diameter.

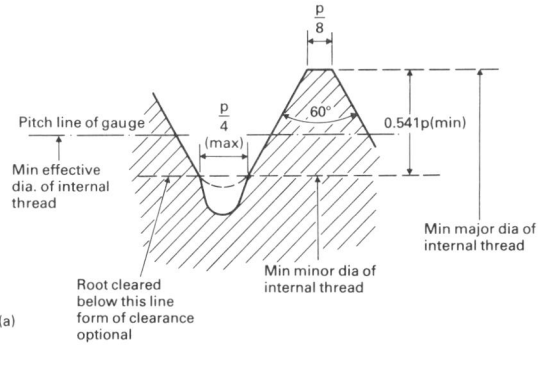

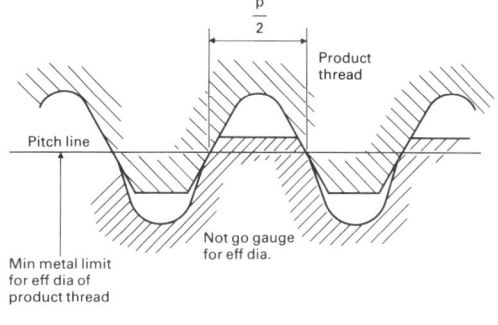

FIGURE 10.15

The root of the gauge is cleared so that it will not be affected by minor diameter errors. The "NO GO" gauge aims specifically to check that the thickness of the thread at the minimum effective diameter is not less than half the pitch. In order to achieve this the gauge has the form shown in Fig. 10.15b. This design clears the gauge from everything but a small length of the flank around the pitch line.

10.8 RING GAUGES

For the inspection of external threads, gauges designed on similar principles but in the form of rings, as shown in Fig. 10.16, are used.

In the case of external threads, the threaded gauge does not check the major diameter. This can be covered by either a direct measuring micrometer or by a plain caliper or snap gauge, as illustrated in Chapter 4, Fig. 4.50.

A weakness of all types of thread gauges is that they wear rapidly, and this has prompted designs which are adjustable.

In the case of ring gauges, this is comparatively easy. The ring is split, as shown in Fig. 10.17, and the gap has a screw across it with which the gap, and consequently the effective size of the ring, is adjusted. Ring gauges of this type require special plug gauges for setting.

FIGURE 10.16

As far as plug gauges are concerned, the problem of achieving adjustability is much more difficult. Although designs have appeared, plug gauges are easier to make, and thus cheaper than ring gauges, and the most reliable and economic course is likely to be to use normal solid gauges and replace them when worn below useable limits.

FIGURE 10.17

10.9 THREAD TOLERANCES

For any particular thread size, the Unified series has three tolerance options defined as Class 1, 2 or 3. The suffix A or B is added to indicate whether the thread is external or internal.

Class 1 is defined as a loose fit which will permit quick and easy assembly, even if the threads are dirty or slightly damaged.

Class 2 is the normal tolerance and permits free assembly, even if this is done by high-speed mechanical methods. It also provides a minimum clearance for the application of a lubricant, when required.

Class 3 provides a close fit and should only be used when exceptionally high performance is required. It demands high-quality equipment for manufacture and inspection, and particular care in assembly.

The class required is included in the designation of the thread. Also included is the number of threads per inch, and it is permissible to specify the size by either a fraction or decimal equivalent. The full definition of, for example, a ¼ UNC and UNF thread might be:

$$\frac{1}{4} - 20 \text{ UNC} - 2A$$
$$0.250 - 28 \text{ UNF} - 3B$$

Figure 10.18 indicates the relationship between the clearances and tolerances for the three classes.

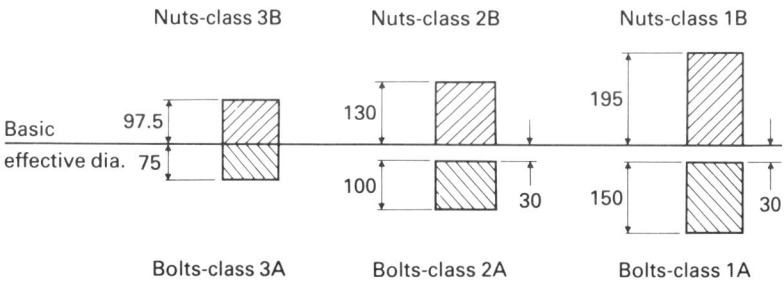

FIGURE 10.18

In this diagram the tolerance on a 2A thread is taken as 100. The tolerance on the effective diameter of a ¼ UNF–2A thread, incidentally, is 0.0033 in.

10.10 COATING OF THREADS

It is quite common for screw threads to be coated, for example by cadmium or silver plate. This is done to provide protection against corrosion, and it also has the effect of lubricating during assembly. The thickness of coating is normally a small fraction of the thread tolerance, and in the case of Class 1 and 2 threads it can be absorbed into the tolerances and clearances without making any special allowance. For Class 3 threads, however, some change will probably have to be made to the machining dimensions to allow for the thickness off coating.

10.11 GAUGE TOLERANCES

The plug and ring gauges used for checking threads must themselves have tolerances. The determination of these is complex, since the gauge must ensure that the thread being checked is within its tolerance, regardless of where the size of the gauge lies within its own tolerance. Recognition has also to be given to the fact that thread gauges do wear quite quickly, and the gauge tolerances have to permit the system to operate economically.

Full details of gauge tolerances are available in various British Standards Institution publications, but Fig. 10.19 shows the tolerance pattern for ⅜ – 24 UNF Class 2 threads.

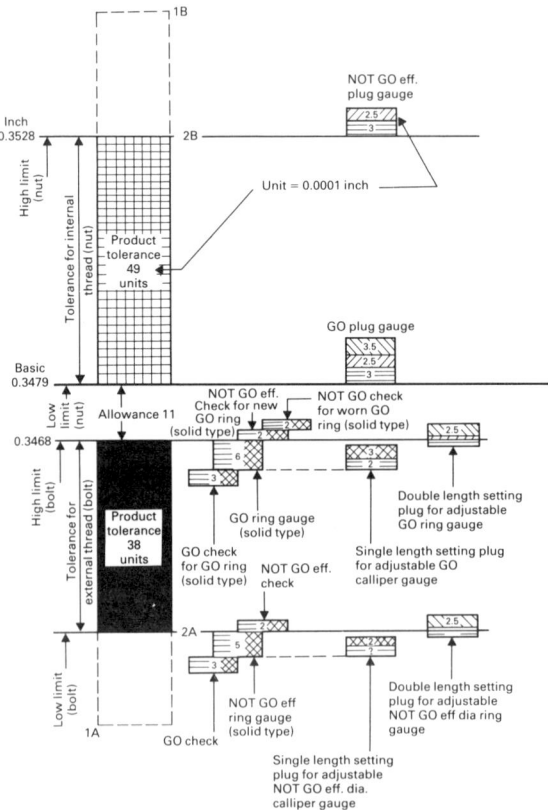

FIGURE 10.19

10.12 OTHER THREAD MEASURING METHODS

As an alternative to ring gauges, a form of gap gauge is available to check external threads. An example is shown in Fig. 10.20.

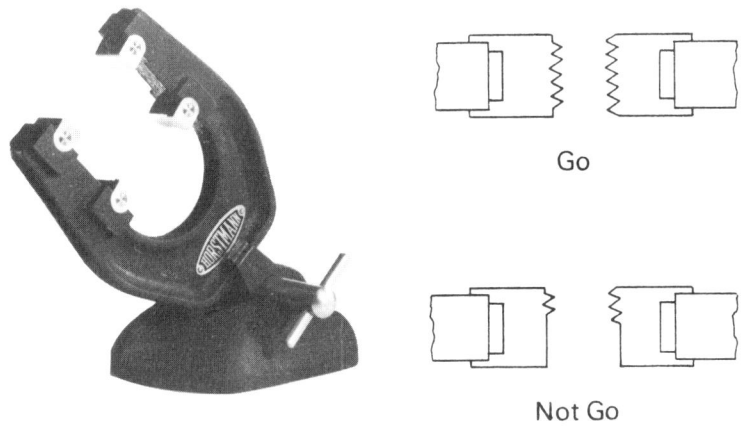

FIGURE 10.20

It consists of a frame in which anvils having the cross-section of a ring gauge are mounted. The "GO" and "NOT GO" threads are mounted in the same frame and the component thread should pass through the "GO" anvils but not the "NOT GO". This instrument is set in a similar manner to the adjustable ring gauge.

Micrometers designed for measuring screw threads will be found amongst the special versions available in the catalogues. They have special anvils, as the example in Fig. 10.21, shaped to check external thread features such as effective and minor diameters.

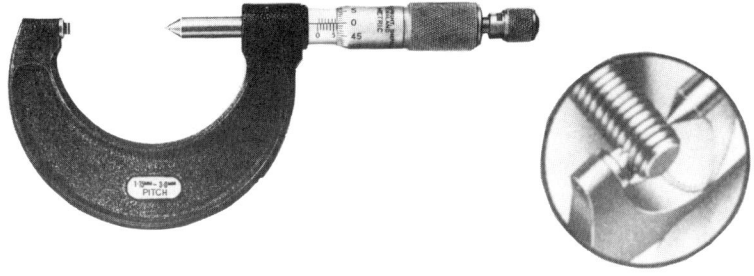

FIGURE 10.21

The anvils are usually interchangeable, so that various features and sizes can be checked with a single micrometer.

Another idea for easing the use of a micrometer in thread measurement is the provision of thread measuring cylinders in the form of a spring, as shown in Fig. 10.22.

FIGURE 10.22

The spring pressure causes the wire to sit tightly in the thread being measured, and the device is particularly valuable for small threads.

Chapter 11

Viewing Devices and Optical Measurement

Although optical means are used to display instrument readings in some of the equipment already described, optical methods may efficiently be used more directly in many metrology and metrology associated applications.

The most obvious is simple visual assessment. This is valuable in the routine evaluation of qualitative parameters such as general finish. It may also be used for other features which, although specified dimensionally, might be adequately judged on appearance. Corner radii are a common example.

11.1 VISUAL ASSESSMENT

The weakness of straightforward visual assessment is that it is subjective, and whether or not a particular feature is acceptable will depend to quite a large extent on personal opinion. The method can be more satisfactorily used if something is done to reduce the amount of subjective judgement involved. An easy way of achieving this is to make the assessment comparative. The comparison is made between the feature being assessed and either a sample of the actual part of a test piece made to simulate the particular features. These samples are measured quantitively by some first principle method to ensure that they are within the specified requirement, preferably somewhere near the middle of the tolerance range.

Clearly a most important factor in making visual assessments is the adequacy of illumination. This is often given insufficient attention and, even for the simplest of visual tasks, it is advisable to specify precisely the intensity of the lighting and its location relative to the workpiece. It is also desirable that regular checks, using a light meter, are carried out to ensure that unnoticed gradual deterioration has not taken place. This is particularly important if fluorescent lighting is being used.

150 Viewing Devices and Optical Measurement

Visual assessment can usually be made more effective by the use of magnification. It does not follow, however, that the higher the magnification the better the result, since high magnification is accompanied by a smaller field of view and shorter depth of focus. Having once selected the most useful magnification, it is important that it is recorded and is always used for this particular task.

Rather more sophisticated devices are available which can make this process more effective. They take the form of magnifiers mounted on stands, often combined with a light source.

A selection of magnifying and illuminating instruments is shown in Fig. 11.1.

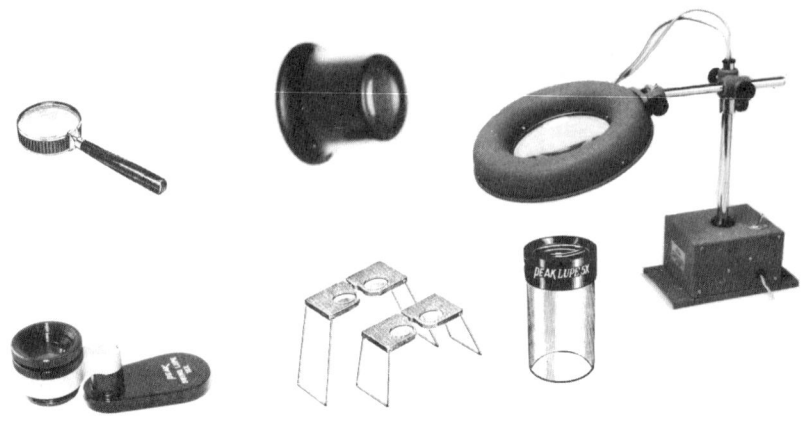

FIGURE 11.1

There are also means for making visual assessment quite critical, generally by enhancing the feature being assessed. An ingenious example is shown in Fig. 11.2, which is for visual assessment of an edge radius. This device emphasises prominently any deviations from a true radius.

A "striped" line of light is projected at an angle on to the edge in question. In principle, if the angle of the light is 45°, then, when viewed perpendicular to the edge, the image received will be equivalent to a cross-section. In practice it is difficult to achieve an angle of 45°, but the approximation if the angle is smaller than this is good enough to make quite an accurate judgement.

Devices of this sort can be made even more critical by using closed circuit television to produce a picture on a screen, as in this example.

Projection 151

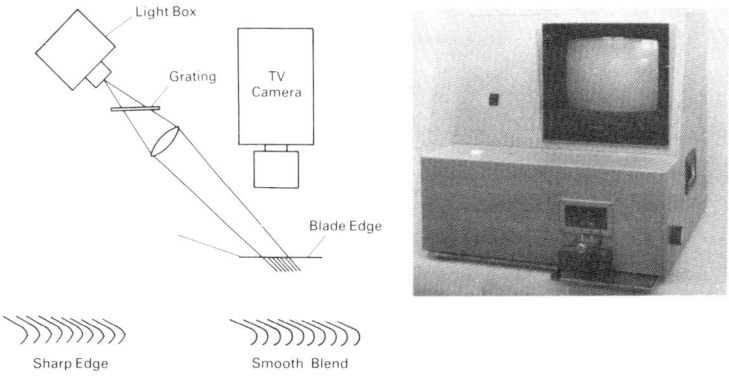

FIGURE 11.2

11.2 PROJECTION

The second, and perhaps most common, use of light as a means of measurement is optical projection. Projectors are available in large variety and some of their uses have already been described in connection with the measurement of contoured surfaces in Chapter 9.

Projectors normally take one of two configurations, which might be described as horizontal and vertical. Examples of both and a diagrammatic indication of their optical systems are shown in Fig. 11.3.

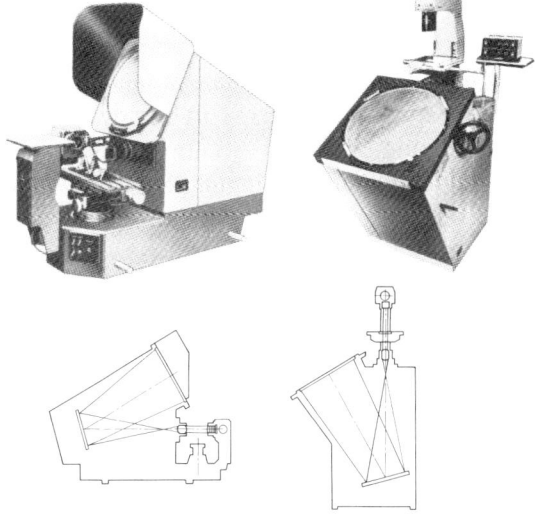

FIGURE 11.3

152 Viewing Devices and Optical Measurement

In addition to its obvious value in assessing profiles of complex or irregular shape, projection can be used as a quick and efficient method of measuring linear dimensions to a limited order of accuracy. This method is particularly suitable for flat parts of thin section. Parts of this nature are quite common and are generally made from sheet metal, as the example of a locking clip in Fig. 11.4.

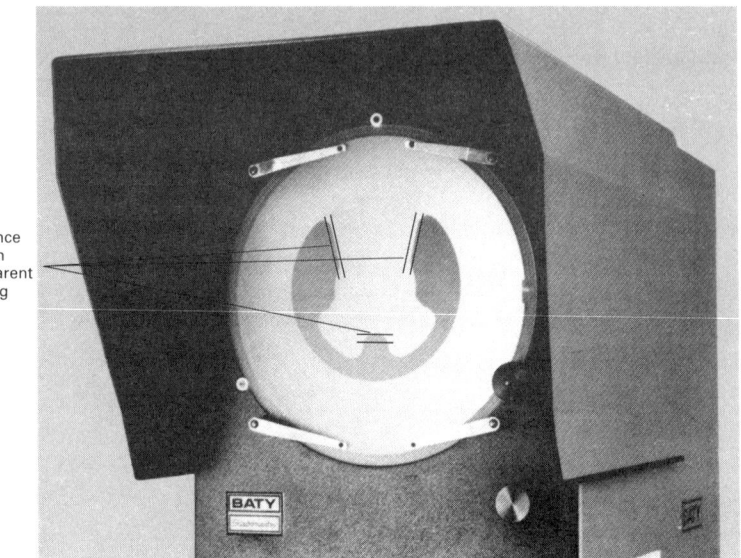

FIGURE 11.4

When using projection for this purpose a number of points must be remembered. One of the most important is that, if the section is of significant thickness, the surface being measured must be precisely parallel to the optical axis of the projector. In Fig. 11.5, if a thick cylinder of actual diameter A is projected with its axis at an angle to the axis of the projector, the image will be of the dimension P which, it will be seen, is greater than A. In addition, the image will be unclear because it is not likely to be possible to have all the projected edges in the focal plane of the lens.

Although measurement can be made simply by comparison with an enlarged transparent drawing of the feature, as in Fig. 11.4, greater accuracy can be obtained by aligning one end of the dimension being measured against a datum, and then moving the part under the control of a linear measuring device until the other end is aligned to the same mark. The dimension is then taken from the reading on the measuring device. This method is depicted in Fig. 11.6.

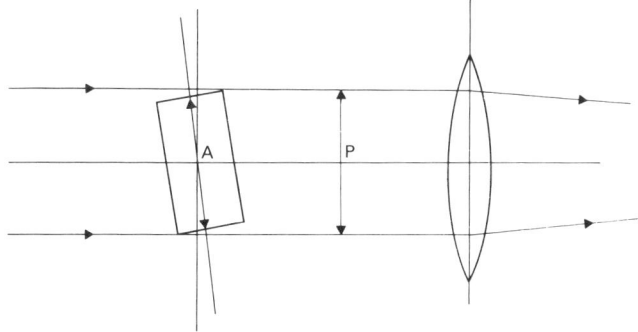

FIGURE 11.5

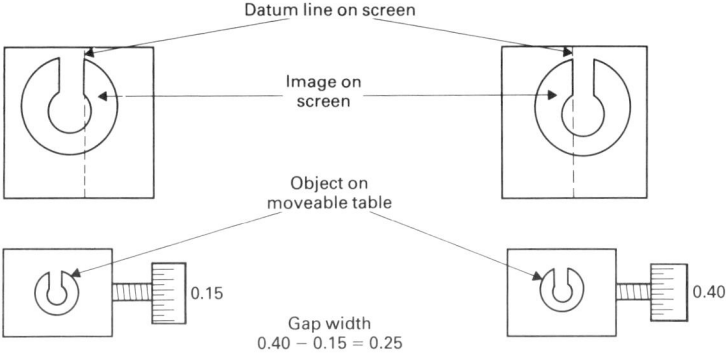

FIGURE 11.6

A sophisticated example of a projector designed to operate in this fashion is shown in Fig. 11.7. This projector has built-in linear measurement with a digital display.

Projectors are, of course, capable of producing images with a wide range of magnification, usually by means of interchangeable lenses. As in the case of simple magnifiers, it must not be assumed that the greater the magnification, the greater the accuracy. The choice of magnification is a compromise between the size of the dimension to be measured, the clarity of the image and the tolerance requirements.

Many projectors are equipped with a facility which illuminates the surface of the part being measured, as shown in Fig. 11.8. This is a useful additional feature, but since the image is produced only by light reflected from the surface of the part, it requires illumination of high intensity. Normally, surface illumination is provided by an additional light source which is reflected on to the surface by a half reflecting mirror. This enables the reflected light to pass through the mirror and produce the screen image.

154 Viewing Devices and Optical Measurement

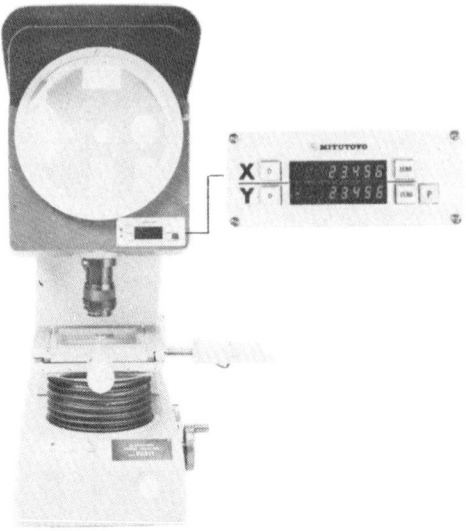

FIGURE 11.7

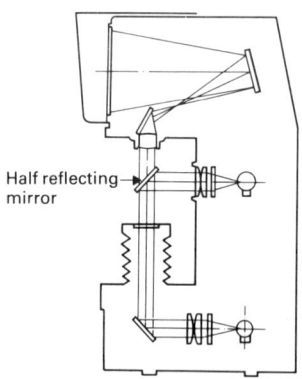

FIGURE 11.8

In this example, surface illumination is greatly assisted by additional local illumination provided by fibre optic light guides.

Although not directly associated with measurement, fibre optics is a use of light which can be of considerable assistance in visual metrology activities. It is a means of conducting light of high intensity along flexible glass fibres with very little loss, and it permits the illumination of areas which might otherwise be inaccessible. Figure 11.9 shows an elaborate viewing device which combines in one cable the fibre optics for both illuminating and viewing the object. It also has built-in variable magnification. Devices such as this were originally developed for the medical profession and, in engineering, are particularly useful for examining the internal condition of a mechanism without the need for taking it apart.

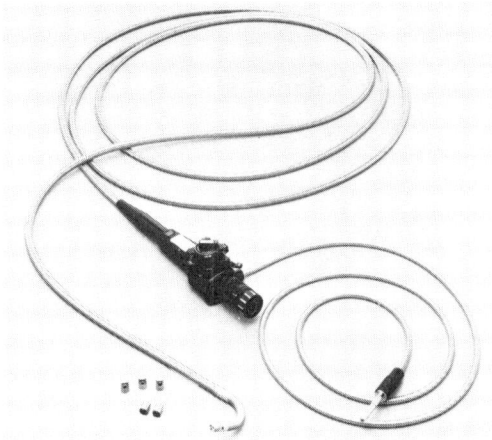

FIGURE 11.9

11.3 NON-CONTACT MEASUREMENT

The concept of measuring an engineering component without any physical contact between the part and the measuring device has many attractions. One non-contact method is by using air gauging, mentioned in Chapter 4, and another is projection. A number of other methods using different scientific principles are likely to develop, but currently projection and non-contact probes using lasers show the greatest potential for achieving the objectives.

The most significant advance in making the projector an instrument capable of accurate non-contact measurement is the introduction of what is effectively a photoelectric eye.

The projector in Fig. 11.10 is designed for this purpose and the photoelectric sensor may be seen in the centre of the screen.

This sensor detects a change between dark and light and is able to define the boundary with remarkable and consistent accuracy. Although the actual value will depend on the sharpness of the image, consistency of 0.01 mm would not be exceptional.

156 Viewing Devices and Optical Measurement

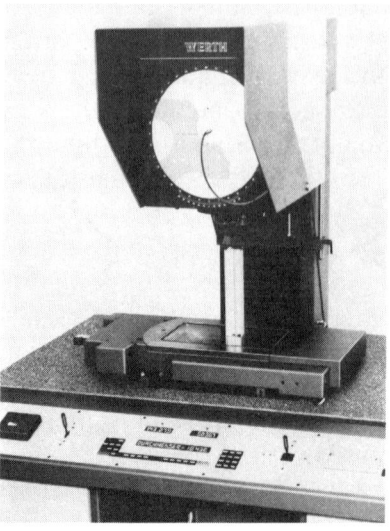

FIGURE 11.10

To make full use of this method, measuring systems producing electrical signals are required on the movements of the worktable. The outputs from both the sensor and the measuring systems are then fed into a computer.

In operation, the worktable is moved manually in an almost random manner, along the lines of the diagrams in Fig. 11.11.

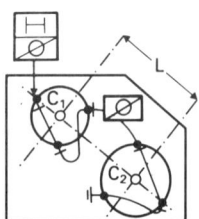

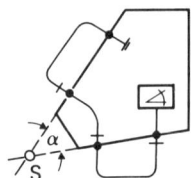
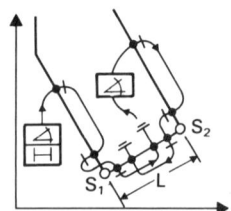

FIGURE 11.11

Non-contact Measurement 157

The only requirement is that the path crosses each straight line feature twice, and any circular features three times.

An important characteristic of the sensor is its ability to trigger the measuring systems accurately whilst the image is moving continously beneath it. It is then possible for all readings to be taken "on the fly" and processed by the computer to give us much detail as is required.

From the readings obtained from following paths as indicated in Fig. 11.11, it is possible to record the positions of straight features relative to each other, either as linear distance or angle between them. The computer will also calculate the diameters of all circular features and the positions of their centres.

A further refinement is that the movements made manually on one component can be made to produce a programme in the computer whilst they are being carried out. If other components of the same type are subsequently inspected, this programme can be used to trace the same path automatically.

There are limitations on the use of measuring machines of this type in that they are only two-dimensional. In fact optical machines are only really suitable for inspecting flat parts of thin section, parts which, in themselves, are effectively two-dimensional.

This limitation can, to some extent, be overcome by the machine shown in Fig. 11.12 which operates on a similar principle but uses a video camera instead of an optical screen.

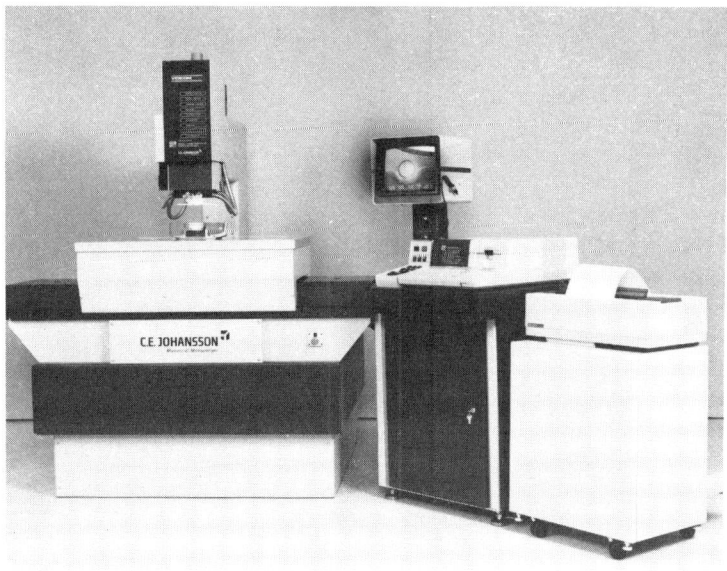

FIGURE 11.12

158 Viewing Devices and Optical Measurement

The video camera has a better capability for detecting surface colour changes and makes the inspection of three-dimensional components, as in Fig. 11.13, practicable.

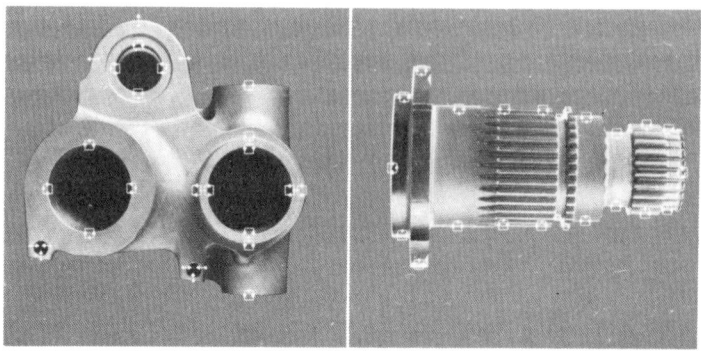

FIGURE 11.13

11.4 TELESCOPES AND COLLIMATORS

A completely different use of light is its application for long distance and angular measurement.

Optical instruments are almost universal for the long distance requirements of surveying, and similar instruments have been adapted for use in engineering applications.

The basic instrument in this context is the alignment telescope, as shown in Fig. 11.14.

It consists essentially of an optical system capable of being focused from zero to infinity. It has a cross wire graticule positioned so that, to the user, it is superimposed on the image being viewed. It also has a simple and accurately produced cylindrical external casing which is parallel to and concentric with the optical axis to a high order of accuracy, better than 3 seconds of arc and 6 µm respectively.

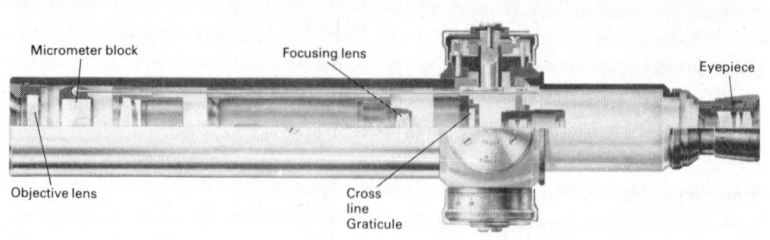

FIGURE 11.14

The telescope is normally mounted in a fixture of the type also shown in Fig. 11.14 and this allows its axis to be adjusted relative to the surface on which the fixture stands.

Other accessories are also necessary and the choice of these will depend on the task to be undertaken. The first requirement is likely to be a simple illuminated target, as the example in Fig. 11.15 shown with a graticule of concentric circles, and the telescope mounted in an adjustable holder.

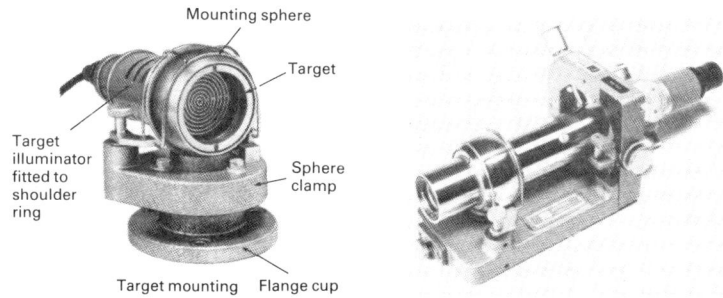

FIGURE 11.15

This combination could be used for checking the flatness of a machine bed or surface table as indicated in Fig. 11.16.

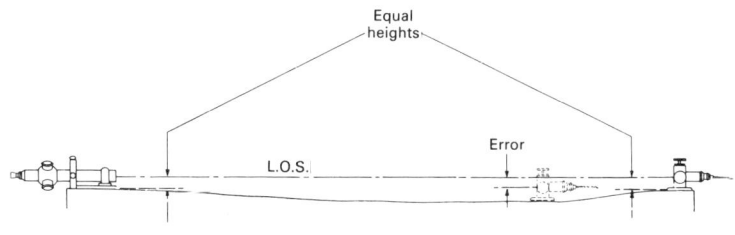

FIGURE 11.16

For such a check, the telescope would be mounted at one end of the surface and the target at the other. The telescope would be focused on the target and the position of either telescope or target adjusted until they are accurately aligned. The target would then be moved to intermediate positions when some displacement might be observed, as in Fig. 11.17, view (a).

This may be evaluated by moving the target until alignment is again obtained and measuring the amount of movement. The telescope in Fig. 11.17, however, is fitted with an optical micrometer which allows the displacement to be read directly.

160 Viewing Devices and Optical Measurement

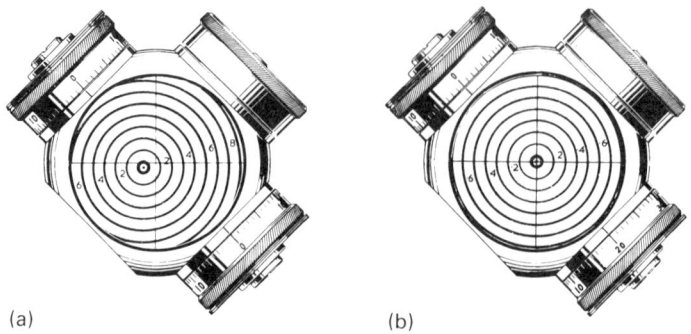

(a) (b)

FIGURE 11.17

The optical micrometer is a thick piece of glass which has flat parallel faces. It may be seen near the objective lens of the telescope in Fig. 11.14 and it can be rotated in two directions at right angles by movement of the micrometer drums located near the eyepiece. If the glass block is rotated so that its faces are no longer perpendicular to the optical axis, refraction occurs which will displace the image with respect to the graticule, as shown in Fig. 11.18. The apparent target image is realigned to the telescope graticule in this way, as in view (b) of Fig. 11.17, and the equivalent linear displacement can be read from the micrometer drum to an accuracy of about 0.05 mm at a distance of 30 m.

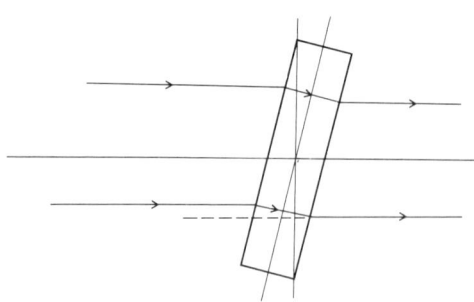

FIGURE 11.18

Perhaps the most complex of the accessories is the collimator. This is shown in Fig. 11.19 together with a cross-section.

Like the telescope, it has an accurately produced cylindrical external form, and its geometrical and optical axes are precisely aligned. It has, at G1, a target graticule which is used in exactly the

Telescopes and Collimators

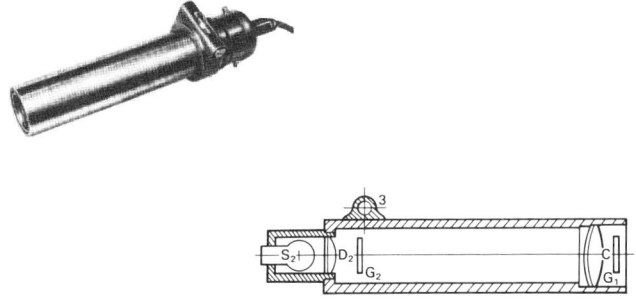

FIGURE 11.19

same way as the simple illuminated target described earlier, and, at G2, another graticule positioned at the focus of its lens, and this graticule is calibrated in angles. In use, if the telescope and collimator are both accurately located in bores a distance apart, the telescope may be focused on graticule G1 and the displacement measured. If the telescope is then focused at infinity, graticule G2 will be seen and its intersection with the cross line of the telescope graticule will show the angle between the bores. The images seen in the telescope eyepiece will be as in Fig. 11.20.

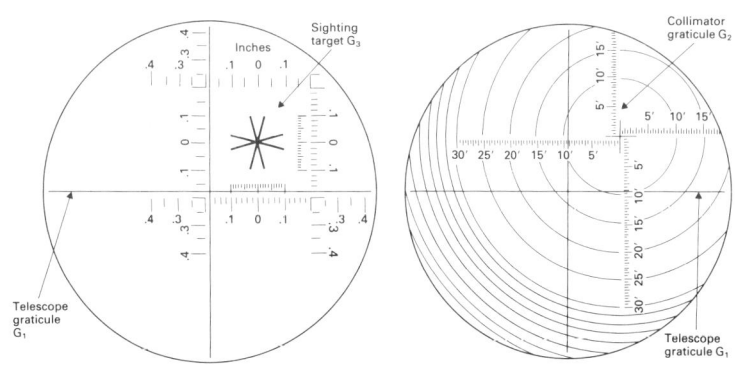

FIGURE 11.20

The target graticule G1 in this case is of the crossed vee type and it also incorporates linear scales. This example indicates that the target is displaced relative to the telescope axis upwards and to the right by 0.018 inches in each plane.

The angle graticule G2 indicates that the axis of the collimator slopes downwards and to the left by 9.25 minutes in both planes.

11.5 THE AUTOCOLLIMATOR

This instrument combines the features of both telescope and collimator, and it is generally more useful than the two separate instruments.

It is shown mounted on an adjustable base in Fig. 11.21, together with a diagram showing the principles of the optical system.

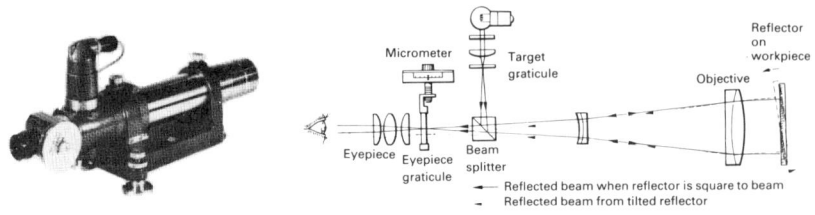

FIGURE 11.21

This instrument is a telescope permanently focused at infinity. If a target graticule is incorporated into the same instrument, it will function in reverse as a collimator. This is achieved by placing an illuminated target graticule in a path at right angles to the main optical axis. The optical system allows both this and the normal eyepiece graticule to be viewed at the same time by turning the light through 90° using a beam splitter, such as a semi-reflecting prism. The light from the "collimator" part of the system is returned to the "telescope" section by a reflecting surface mounted on the subject of the particular measurement. The only measurement the instrument makes is the angle between its axis and the reflecting surface. If this angle is 90°, the eyepiece and target images will coincide, and any displacement will be proportional to the angular difference. This can be quantified by moving the eyepiece graticule, under the control of a micrometer, until the images again coincide. An example of the view through the eyepiece is shown in Fig. 11.22.

The micrometer drum in rotated until the double lines on the eyepiece graticule straddle the single line on the reflected target. The angle can then be read to an accuracy of around 0.2 seconds of arc.

Although the ability to measure angular deviations in this way may, at first sight, appear to be of limited value, the capability of measurement of small angles to a very high level of precision makes the autocollimator a very versatile instrument.

The Autocollimator 163

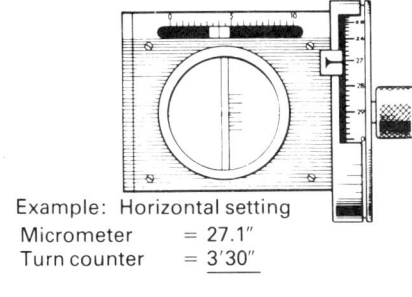

Example: Horizontal setting
Micrometer = 27.1"
Turn counter = 3'30"

Reading = 3'57.1"

FIGURE 11.22

Figure 11.23 shows an example of fairly obvious use in checking the flatness of long bed, and also an example of a more ingenious application.

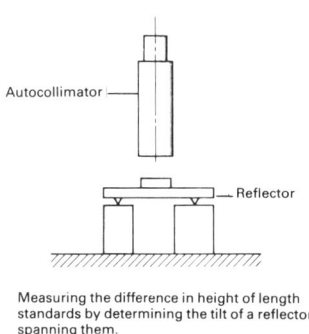

Measuring the difference in height of length standards by determining the tilt of a reflector spanning them.

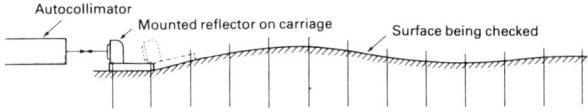

Measuring the departure from flatness using a reflector carriage

FIGURE 11.23

This checks the differences between the two length standards by measuring the angle between the surface on which they stand, and a surface spanning them.

Autocollimators are available in varying degrees of sophistication. The simplest is known as the Angle Dekkor and is shown in Fig. 11.24.

164 Viewing Devices and Optical Measurement

FIGURE 11.24

The readings of this instrument are obtained directly from a scale in the eyepiece graticule, to an accuracy of about 6 seconds of arc.

At the other end of the scale is the photoelectric autocollimator shown in Fig. 11.25. This uses photoelectric means to align the target and eyepiece graticules and also gives a digital readout of the angle. It is capable of consistent accuracy to 0.5 seconds.

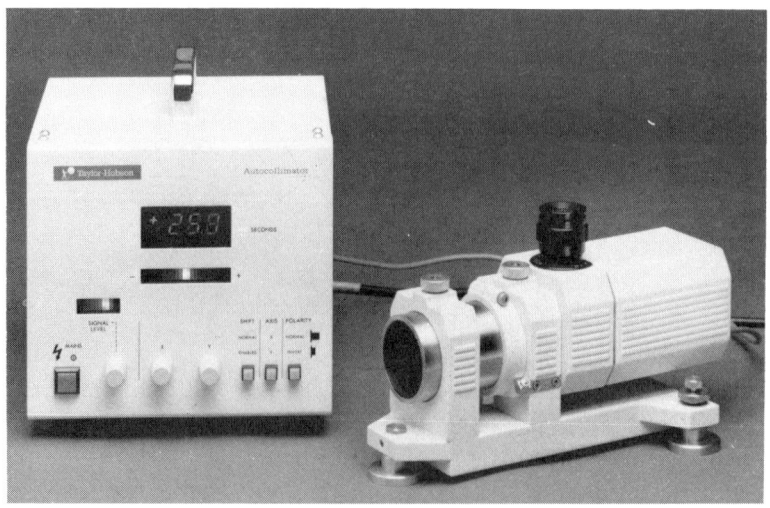

FIGURE 11.25

11.6 INTERFEROMETRY

Interferometry is a form of measurement which makes use of the wave form properties of light.

Interference occurs when two light beams of the same wavelength meet. Figure 11.26 indicates two rays of light, A and B, of the same wavelength but unequal amplitude which are "in phase".

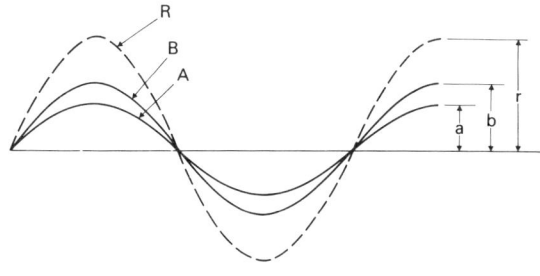

FIGURE 11.26

The visible result will be R, the sum of the two rays. If, however, the rays are out of phase by 180°, the situation will be as seen in Fig. 11.27.

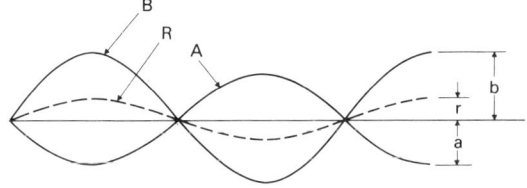

FIGURE 11.27

In this case, the visible result R will be the difference between A and B. If the rays are completely monochromatic and of equal amplitude, the result in the Fig. 11.26 case will be visible light double the brightness of a single source, and in the Fig. 11.27 case the result will be darkness. If normal white light, which has a combination of many different wavelengths, is used, the phe-

166 Viewing Devices and Optical Measurement

nomenon is apparent as a rainbow appearance. This does not completely preclude its use for measurement purposes, but monochromatic light, which is fairly easy to produce, is much more satisfactory.

The basic piece of equipment required to make use of this principle is very simple indeed, consisting of a piece of glass of substantial thickness with a very accurately polished flat surface. Such a piece of glass is known as an optical flat. If this is placed close to, but inclined at a small angle with, a reflecting surface, as indicated in Fig. 11.28, a ray of monochromatic light striking the upper surface of the flat at A will be bent due to refraction at this surface and then strike the lower surface at B.

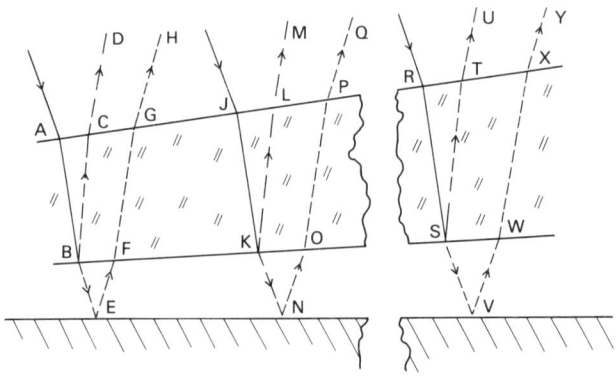

FIGURE 11.28

At this point, part of the light will be reflected internally in the optical flat and will follow a path BCD. The remainder will, after further refraction, follow a path BE, striking the reflecting surface at E, and return to strike the flat again at F. This ray will then follow the path FGH and an observer at D and H will see a combination of both rays.

The additional path length BEF followed by the second part of the ray causes a phase difference between the two parts, and if this difference is equal to an odd number of half wavelengths of the light, the visible result will be darkness. Conversely, if the path difference is equal to an even number of half wavelengths, the visible result will be bright.

Since the optical flat is inclined to the reflecting surface, the difference in path length of the parts of the rays will vary. In Fig. 11.28, KNO and SVW are both longer than BEF. Thus, if the total top surface of the flat is observed, and the reflecting surface itself is

Interferometry

flat, the appearance will be a succession of light and dark bands, as depicted in Fig. 11.29.

The number of bands visible is not a means of measuring flatness, since it depends entirely on the angle between the surfaces. Figure 11.29 (b) is the same surface as Fig. 11.29 (a) with a greater angle of inclination. As the angle becomes larger, the number of bands increases, but, if the surfaces are perfectly flat, they will always be straight and equally spaced.

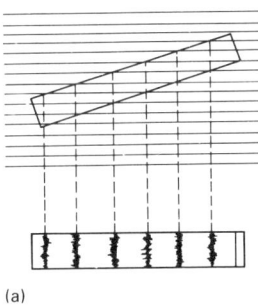
(a)

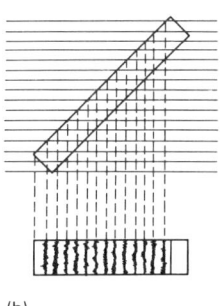
(b)

FIGURE 11.29

The interference "fringes", as these bands are often called, may, in fact, be interpreted as contour lines and they give a measure of flatness in this way.

Figure 11.30 gives an indication of how a cylindrical surface might appear with the optical flat angled perpendicular to and along the cylindrial axis.

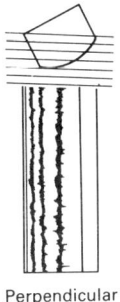
Perpendicular

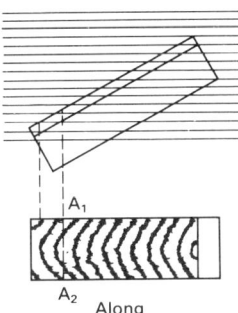
Along

FIGURE 11.30

168 Viewing Devices and Optical Measurement

This figure also illustrates the way in which flatness errors can be calculated. If a line is drawn between the points A1 and A2, it will be seen that the deviation from straight of fringe A is 1.5 wavelengths of the light used. Since this is likely to be of the order of 0.0008 mm, this represents a flatness error of 0.0012 mm, which is an indication of the sensitivity of the method.

In practice, adequate results can be obtained merely by placing the optical flat on the surface to be checked. The angle between the surfaces required to produce fringes is extremely small, and it will normally be created by the air between them. Figure 11.31 shows patterns obtained by placing an optical flat on a collection of gauge blocks.

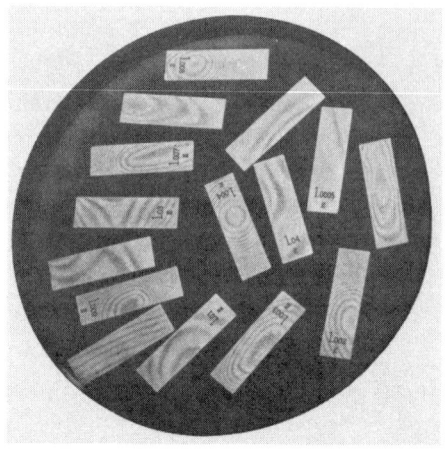

FIGURE 11.31

Figure 11.32 shows another simple use of the optical flat, for checking the flatness and parallelism of micrometer anvils, together with the flats used for this purpose.

These "flats" have both their surfaces very accurately parallel to each other and, for this reason, are knonw as "optical parallels". They are supplied in sets of four, which differ in thickness by an amount represented by one-fifth of a revolution of the micrometer spindle. This enables the anvil faces to be checked at four different angular positions.

Whilst optical flats can be used quite successfully under white light, very much better results can be obtained with monochromatic light, and a simple, but effective, monochromatic light source is shown in Fig. 11.33.

Interferometry 169

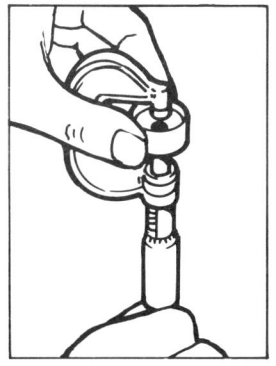

FIGURE 11.32

FIGURE 11.33

The principle of interferometry can be more widely used if the optical flat is incorporated into an instrument, known as an interferometer, of the type shown in Fig. 11.34.

This instrument enables very accurate control over the angle of the flat and this, together with the facility for examining the fringes through a specially designed optical system, makes it possible to check the flatness of surfaces which do not permit the direct application of the flat to the surface.

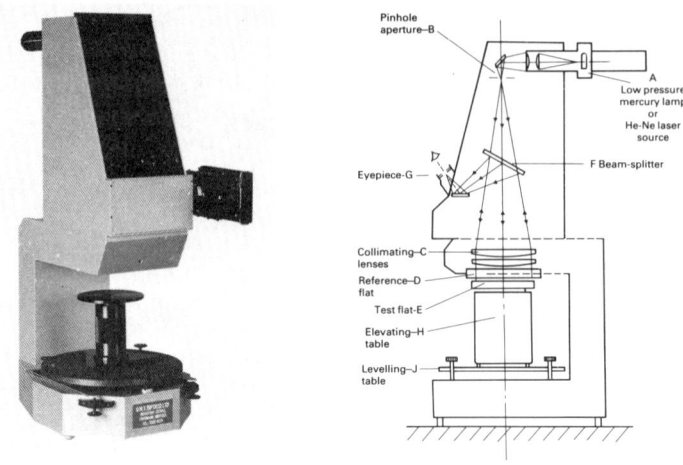

FIGURE 11.34

11.7 LASER

One of the most recent technological developments, which has rapidly increasing application to metrology, is laser light.

Perhaps the most obvious application of the laser is in the field of alignment. Laser equipment is capable of undertaking most of the functions for which alignment telescopes and collimators are used. It would normally be capable of a higher order of accuracy over much longer distances, and its operation would be likely to be more automatic. It is, however, expensive and it is likely that the greater cost would only be justified in specialised applications. A selection of laser equipment used for alignment testing is shown in Fig. 11.35.

Rapid developments are taking place with regard to the use of lasers in other spheres, particularly non-contact gauging. An example is shown in Fig. 11.36 of an instrument which makes use of the precision of the laser beam to cast a measurable shadow of an object placed in its path.

A thin band of laser light is projected from the transmitter on the right, and the receiver on the left is able to determine the diameter of the bar by measuring the width of the shadow. An instrument such as this can be read to 1 micron with a repeatability of 5 microns.

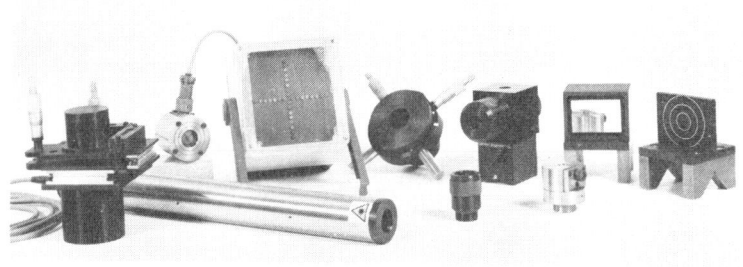

FIGURE 11.35

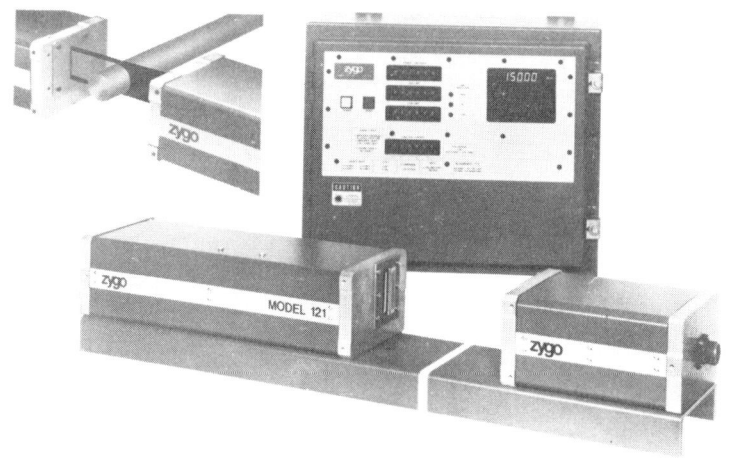

FIGURE 11.36

A variety of non-contact probes are now becoming available which use laser light in a less direct way, as the example in Fig. 11.37.

A laser beam is focused on the surface to be measured and the light scattered from the surface strikes two detectors. The distance from the probe to the surface is measured by the position of the projected light on the detectors. This probe will operate accurately with a distance between probe head and surface of up to 50 mm.

A laser version of the probes discussed in Chapter 5 is shown in Fig. 11.38.

172 Viewing Devices and Optical Measurement

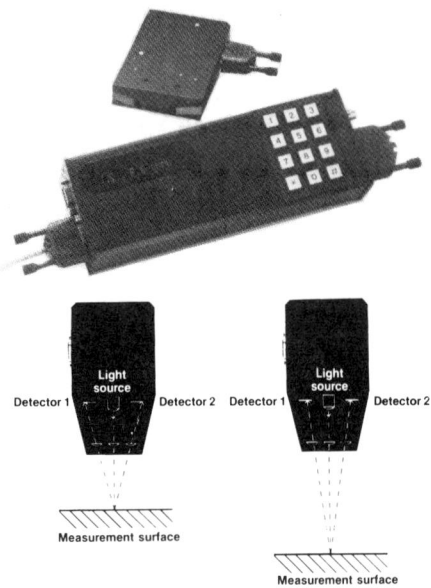

FIGURE 11.37

FIGURE 11.38

Although much more expensive than the contact probe, it operates at a distance of 20 mm from the surface, has a measuring range of 4 mm with an accuracy of 10 microns, and it is capable of scanning a surface at a rate of 50 readings per second.

Other examples of the use of the laser make use of the wavelength of the light as the reference scale, and an instrument using this principle is shown in Fig. 11.39.

This is, in effect, a very refined form of height gauge which has a very low contact load and accuracy of around 5 microns per inch.

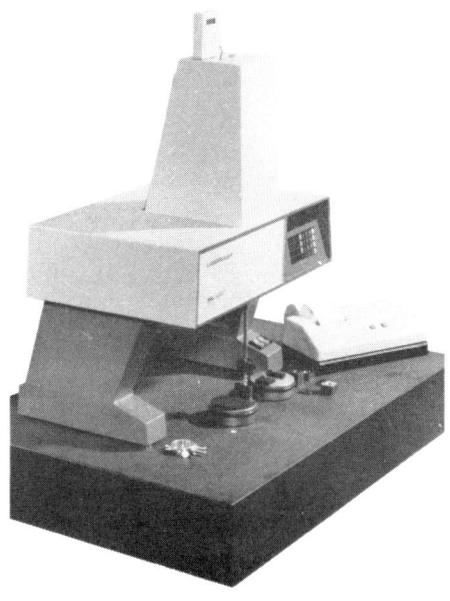

Figure 11.39

Chapter 12

Electronics, Computers and Other Metrology Techniques

This chapter is devoted to equipment and techniques not primarily developed for metrology. Prominent amongst these is the computer, but mention will also be made of some metallurgical procedures which can be of value in dimensional measurement.

12.1 COMPUTERS

The use of computers has already been mentioned several times in connection with a variety of measuring equipment, and it might be useful to summarise their application in the field of metrology.

They serve to main purposes:
– controlling the function of a piece of measuring equipment.
– processing the output of an inspection operation.

12.2 FUNCTION CONTROL

The use of a computer for any purpose necessitates a "programme". This may be specifically written for a particular application, or it may be obtained from commercially available "software". The writing of programmes requires a particular, and probably quite expensive, skill, but the results could be more efficient if the programme is to be frequently used. On the other hand, most manufacturers of measuring machines suitable for computer control also offer a specific computer, and a range of general purpose software to cater for the activities most likely to benefit from control in this way.

In order to make efficient use of computer control, a measuring machine must have movements which are motorised, and it must have a measuring system linked to the movement motors in such a way that the movement will come to rest in a precise predetermined position.

It will be possible for any machine satisfying these requirements to operate under computer control, and it will be possible to use a wide range of computers for this purpose. Programmes are, however, rarely interchangeable between computers, and, if a machine manufacturer offers software, it is almost certain that it will only operate with the computer he supplies or recommends.

Given a suitable measuring machine, a compatible computer, and software to carry out the required task, all that is likely to be necessary is for the operator to carry out instructions the software will cause to appear on the visible display system, possibly a printer but, more likely, a TV monitor screen.

Functional control by computer can be either semi- or fully automatic.

If it is semi-automatic, the machine will stop after completing part of the operation. At this point, it may be necessary for the operator to make some contribution, although this may be no more than pressing a key to trigger the next part of the cycle.

If the operation is fully automatic, when once initiated, the programme will run the machine until the task is complete. This could involve the taking of literally hundreds of readings, analysing them, and producing hard copy of the results. Such an operation may take hours, but the machine will run entirely unattended. Figure 12.1 shows a measuring machine set up with a universal probe to carry out an automatic inspection of this type.

FIGURE 12.1

Another way in which a measuring machine may be programmed to carry out an automatic inspection is to use a computer programmed to compile an operating programme from the procedure used when a sample component is being measured manually. With this technique, the procedure followed whilst measuring the sample will be automatically followed for subsequent components.

12.3 OUTPUT PROCESSING

The potential for using a computer to process measuring machine output is virtually unlimited. In fact its use is not confined to the output from elaborate and expensive measuring machines. Many of the digitised basic instruments already described have facilities for feeding outputs into computers. Figure 12.2 shows a collection of such instruments, all of which feed into the same processing system.

FIGURE 12.2

The collecting device in this case is the portable, hand-held computer shown in Fig. 12.3.

Whilst this has the capability of storing and processing the input it obtains from the measuring instruments, it has simple provision for feeding its contents into the large capacity computer in Fig. 12.4.

This system is designed to collect large quantities of data from different inspection locations in a plant and to analyse it collectively.

Output Processing 177

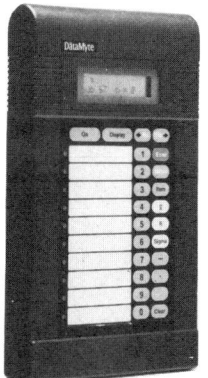 

FIGURE 12.3

FIGURE 12.4

After the obvious capability for determining whether a measured dimension is within required limits, the ability of the computer to store and analyse data obtained from measuring a large number of pieces is perhaps the next most useful feature. This may be presented as numerical tabulations, or in a variety of graphical forms, as the example in Fig. 12.5.

Even more spectacular are three dimensional pictorial representations, as the example in Chapter 9, Fig. 9.15. A large number of parallel surface finish traces processed in a similar way are shown in Fig. 12.6.

178 Electronics, Computers and other Metrology Techniques

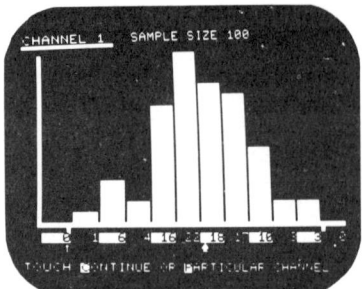 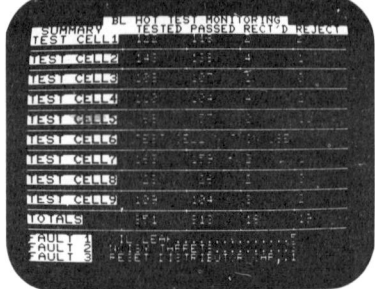

FIGURE 12.5

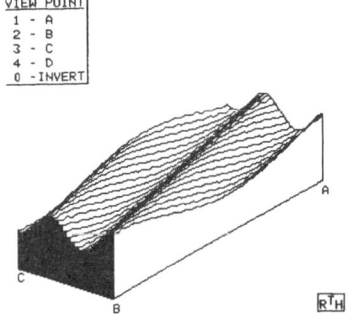 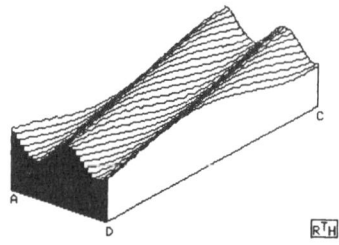

FIGURE 12.6

The computer program for producing this output is able to present the same surface as it would be when viewed from all the corners, and two of the aspects are shown in Fig. 12.6.

Whilst this form of output is very impressive, its interpretation can be more difficult than the more straightforward presentations.

12.4 COMPUTER TYPES

Many manufacturers make use of standard commercially available personal microcomputers, as the example in Fig. 12.7.

These have rather complex keyboards of typewriter layout which might look somewhat forbidding to the uninitiated, and require a certain amount of skill and training to use.

On the other hand, they are able to perform all the normal functions of a microcomputer.

The alternative is an abbreviated computer, or, at least, a computer with an abbreviated keyboard. This can be as simple as a calculator keyboard with a few additional function buttons, as shown in Fig. 12.8.

Computer Types 179

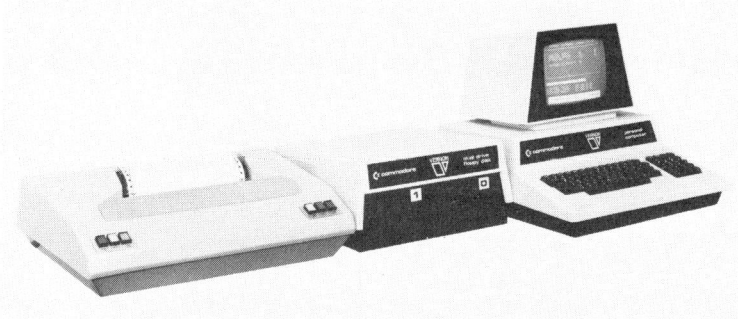

Figure 12.7

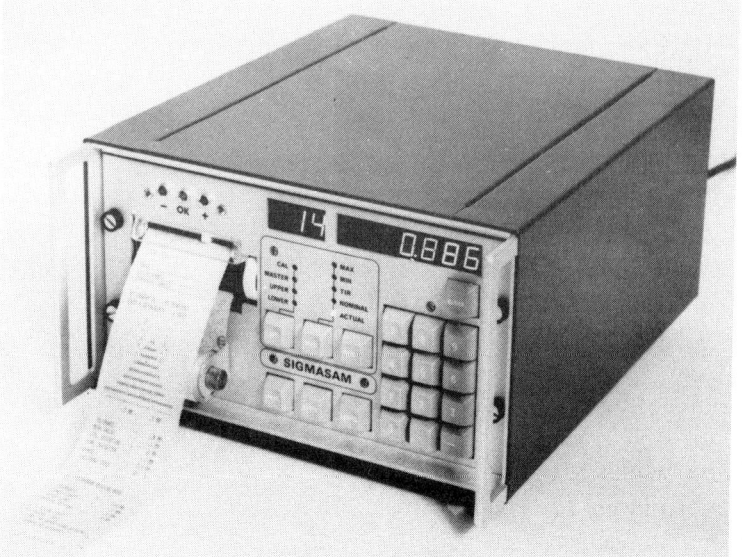

Figure 12.8

A rather more comprehensive, but still specially tailored, keyboard is shown in Fig. 12.9.

One feature of computers of this type, which are tailored to carry out a limited number of functions, is that the same philosophy can be applied to deal with the software. Since the programs required will be limited, they can be permanently incorporated into the computer, and this avoids the need for any external software and the necessary equipment to read it.

180   **Electronics, Computers and other Metrology Techniques**

FIGURE 12.9

The development of special purpose computers and, particularly, calculating devices with limited printing and computing facilities, is rapid, and two applications of this type are shown in Figs. 12.10 and 12.11.

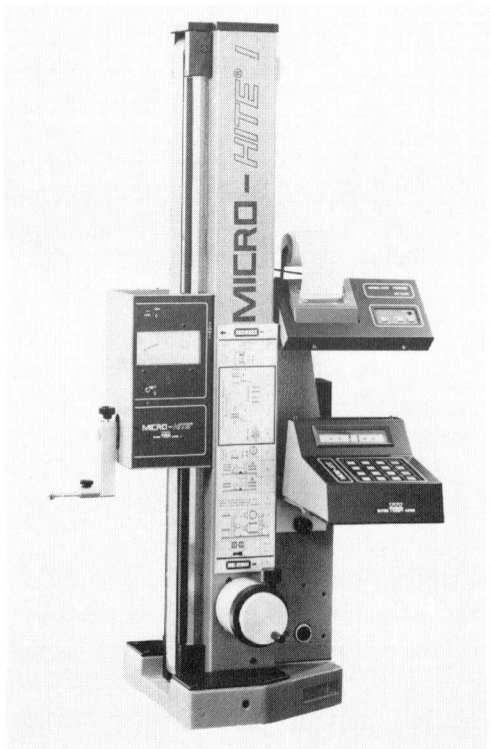

FIGURE 12.10

Figure 12.10 is a development of a height gauge of the type shown in Chapter 5, Fig. 5.3, to include an integral computer and printer.

Figure 12.11 shows an ingenious computer and printer built into a gauge block container.

FIGURE 12.11

Chapter 13 details the method of building up gauge block combinations to give a specific dimension. The computer in this set not only details the sizes required for the dimension entered, it also illuminates arrows above the actual blocks required. The computer has a number of additional facilities, including storage in its memory of deviations from the precise size of all the blocks, printing the total deviation from the combination of blocks for a required size, and storing and displaying information regarding calibration requirements.

The ability of the computer to carry out, almost instantaneously, large numbers of tedious calculations is one which can be exploited to the full.

An example in respect of diameters and positions of circles is described in Chapter 11. Traditionally, the diameter and position of the centre of a hole would be obtained, somewhat painstakingly, by measuring two positions at opposite ends of a diameter. A circle, however, is completely defined by three points on its circumference and, whilst obtaining the required information from any three points is a complex calculation, it is no problem to the computer.

182 Electronics, Computers and other Metrology Techniques

In Chapter 8, various gear tooth measuring machines were described which compare the tooth profile with an involute form generated by the machine itself, basically by rolling a base circle against a straight edge. As a final example of a computer application, the machine in Fig. 12.12 is a gear tooth measuring machine, but the profile against which the part is checked is calculated by a computer within the machine.

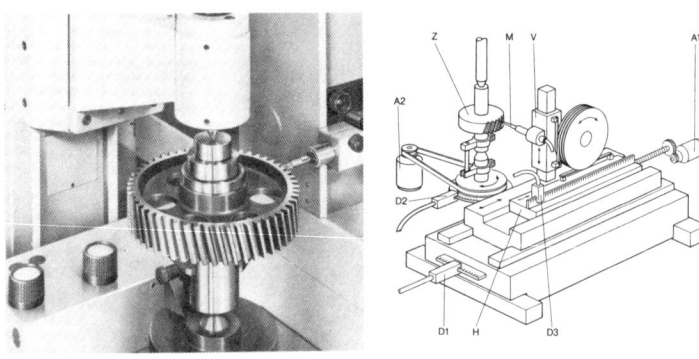

FIGURE 12.12

In this machine, the required base circle diameter is set on the measuring system D1. The gear under test Z is rigidly connected to angular measuring system D2 and this controls the rotation of the gear through motor A2. The carriage H is moved in a direction tangential to the gear by motor A1 and its movement is monitored by the measuring system D3. The measuring stylus M is mounted on a vertical slide V which is attached to carriage H. The outputs from D1 and D3 are fed into the computer, which calculates the required angular movement of the measuring system D2 to ensure that the contact point between the probe and the tooth profile traces a perfect involute.

This is a particularly good example of the potential of the computer, since, although the electronics to achieve this result are at present expensive, the cost is rapidly reducing to a level comparable with the cost of the meticulous manufacture of the parts necessary to produce the same results mechanically.

A situation can thus be foreseen in which electronics, via the computer, will achieve more accurate results than the highest quality mechanisms and at substantially lower cost.

12.5 NON-DESTRUCTIVE TESTING

Non-destructive testing is normally associated with the examination of components for material flaws. As the name implies, it consists of a wide variety of techniques for inspecting metallurgical quality, externally and internally, without destroying the component.

There are circumstances when these methods can be usefully employed for dimensional checking, particularly for features which might be inaccessible to normal metrology equipment.

12.6 RADIOGRAPHY

X-rays are of particular value in revealing internal details and rely for their effect on the difference in density between the base material and the flaw or other feature requiring detection or examination.

They will indicate the presence of changes in section and may be used to verify the position of internal passages and cavities, intentional or otherwise.

When used in this way, the interpretation of an X-ray image is similar to that of a projection. There are, however, two major differences.

First, although so-called "real time", or direct viewing, radiography is possible, current techniques normally involve producing and processing a film. The film is invariably a negative and this makes comparison with a nominal size layout fairly easy.

The second difference, however, is that, because a normal X-ray tube produces rays from a comparatively large area, between 1 mm and 3 mm diameter, pictures are normally taken with the film in contact with the part to obtain adequate definition, and the negative does not have any magnification. This, coupled with the fact that X-ray images are not usually very sharp, severely limits the accuracy of the method.

This situation can be improved very considerably by the technique of high definition radiography. An example of the equipment which might be used, and a diagrammatic representation of the method are shown in Fig. 12.13.

The most important feature of this equipment is that the X-ray source is very small, around 0.05 mm diameter, and, for this reason, the technique is also referred to as microfocus X-ray. This source is effectively a point, which makes it possible to position the film some

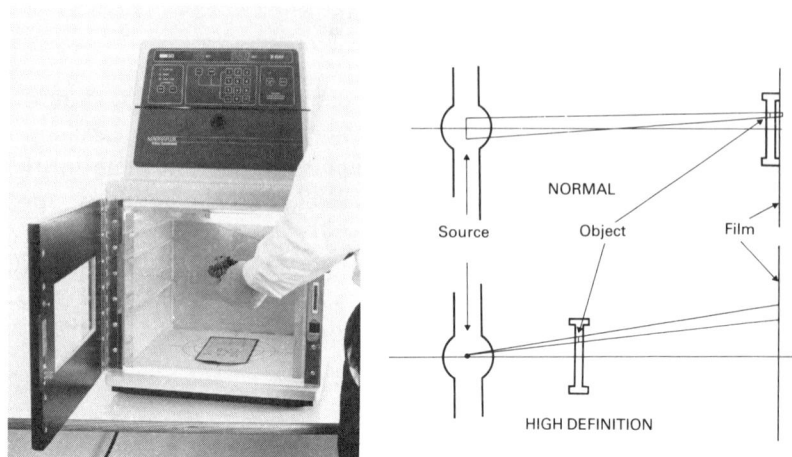

FIGURE 12.13

distance from the object, thus obtaining magnification. Figure 12.14 shows a print of a small part containing a number of internal holes.

The magnification is about ×10, and it will be seen that the definition is sharp enough to measure the section between the holes with an adequate accuracy for many applications.

As will be seen from Fig. 12.13, another advantage of this form of X-ray is that the power is sufficiently low not to require expensive safety precautions, although this does mean that the section thickness through which it can be used is limited.

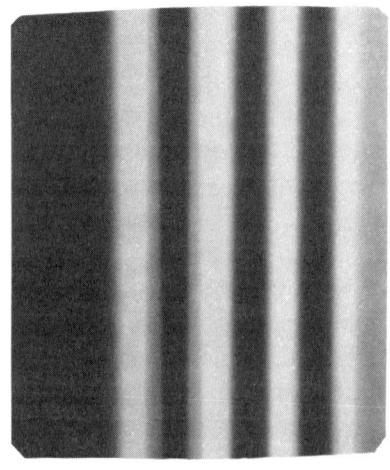

FIGURE 12.14

12.7 ULTRASONICS

This is a technique which transmits ultrasonic sound waves, above 16 kc/s, into a material. A receiver is also positioned on the material, often incorporated in the same head as the transmitter, and the basic data obtained is the time taken for the sound waves to travel from the transmitter to the receiver.

This becomes of value because the sound waves are reflected from discontinuities in the material, including its external surfaces.

The time is actually measured on a comparative basis in that the time from one discontinuity is compared with the time from another. Figure 12.15 illustrates the situation diagrammatically.

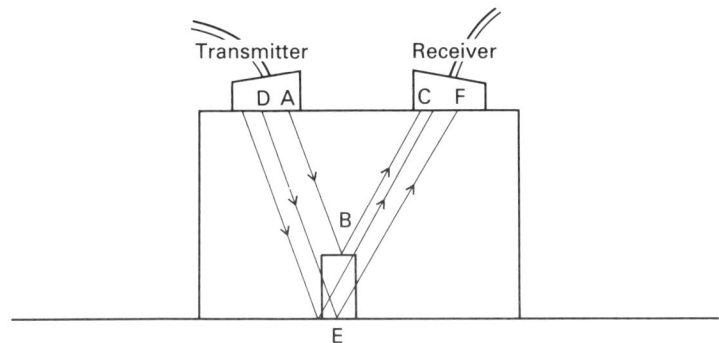

FIGURE 12.15

A wave reflected from the end of a hole in a block will follow a path ABC which is shorter, and faster than a wave reflected from the lower surface following a path DEF.

The normal means of measuring the time is a trace on a cathode ray oscilloscope but this can also be reproduced on a pen recorder. In either case the visual form is the same. Figure 12.16 shows the type of recording which would be obtained from a plain rectangular piece of material, and from the same piece of material containing a blind hole.

The "blips" on the recording appear at positions corresponding to the surfaces of the block and the bottom of the hole. If the total thickness of the block is known, the depth of the hole can be assessed as the proportion BC/AB.

As with most methods of this type, the results are not particularly accurate but can be valuable in the case of parts which when completed, have inaccessible features. Parts which involve fabrication and welding are examples.

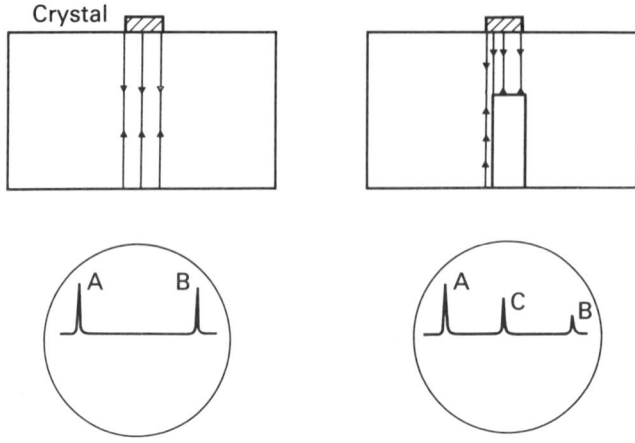

FIGURE 12.16

12.8 GAMMA RADIATION

Although not so widely applicable or so accurate as ultrasonic sound waves, the penetrating γ emissions from radioactive isotopes can be used in a similar way. In this case the method depends on the absorption of the radiation by the material concerned, and, provided a datum can be established, material thickness to a particular feature can be assessed from the degree of attenuation of the signal.

12.9 EDDY CURRENT METHODS

Certain other non-destructive testing methods can be useful for identification of physical characteristics.

An example is the instrument shown in Fig. 12.17 which measures the reaction of a small part to eddy currents.

This is an exceptionally versatile piece of equipment because eddy current response is affected by a wide variety of features. These include material specification, material condition, material defects and geometrical shape.

The instrument in Fig. 11.17 consists of two identical coils as the input and an oscilloscope trace as the output. This trace may be used to compare the contents of the two coils.

In practical use, an acceptable sample of the part to be checked is placed in one of the coils, and the batch is then fed, one by one, into the other. The shape of the trace will then indicate whether or not

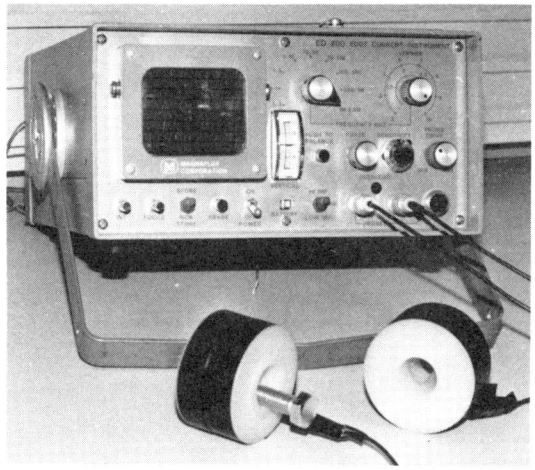

FIGURE 12.17

there is a difference between the two parts. This might be a metallurgical feature, but the instrument reacts well to gross dimensional differences such as a short thread on a bolt.

This instrument can be readily automated, which makes it suitable for high speed "right or wrong" type sorting of small items made in large quantities.

Chapter 13

Standards Room, Laboratory and Special Purpose Measurement

It is highly desirable that any organisation manufacturing an engineering product has a dedicated facility, however small, with the specific task of keeping under control all measurement standards and measuring equipment. The measurement carried out in this facility will be to an order of accuracy higher than elsewhere in the plant and, for this reason, it is preferable for it to be protected, normally by enclosure, from the hazards of the normal shop floor. A facility such as this is generally described as a standards room, or a metrology laboratory.

13.1 OBJECTIVES

The main objective of standards room measurement is to ensure that all measuring equipment is capable of measuring the work to the required order of accuracy and against an adequately qualified datum. Ideally, all engineering measurement, regardless of where it is carried out, should be capable of being related to the national, if not international standards referred to in Chapter 1. At first sight this might appear to be a very difficult and extremely expensive objective to achieve. The facilities available for achieving it, however, are such that even the smallest organisation can economically have its own formally related datum standards and also the means of ensuring that all measuring devises in use in the organisation are regularly checked against these standards.

13.2 INSTRUMENT CONTROL PROCEDURE

In addition to the means for checking inspection instruments to the appropriate level of accuracy, it is essential that a system exists which, as automatically as possible, subjects all such instruments and gauges to regular, and recorded, checks.

The first requirement for the system is a determination of the frequency with which each instrument should be checked. Obviously, if the system is to be efficient, this frequency will vary between instruments of different types, and will depend on the accuracy of the inspection checks the instrument is required to perform, the vulnerability of the instrument to wear, and its normal frequency of use.

At the commencement of the system, these factors can be used to make an assessment of the shortest period during which there is a possibility of the particular instrument or gauge deteriorating to the point at which its accuracy is not adequate for its task. Subsequently, the period between checks can be adjusted in the light of experience. If, for example, an instrument is frequently found to be deficient, the period is too long. On the other hand, if a number of successive checks show no significant difference from previously recorded figures, the period is too short.

The other important requirement for the system is that the user of the instrument is aware that it is within its check period. The most obvious means of achiveing this is for the instrument to be marked, in some semi-permanent manner, with the date on which it is next due for checking. A popular alternative is colour code system which changes at, say, monthly intervals.

13.3 ACCURACY REQUIREMENTS

Having organised the availability of instruments and gauges for checking, the next requirement is that the checks are carried out to the appropriate level of accuracy. In the same way that in Chapter 2 it is noted that measuring equipment should have an accuracy of the order of 10% of the tolerance being measured, the instruments used to check the measuring equipment should have an accuracy of about 10% of the capability of the equipment being measured. As an example, if the concern is ultimately with a component dimension having a tolerance of 0.025 mm, equipment used to measure this should be capable of consistent measurement to 0.0025 mm. In turn, the instruments and standards used to check this measuring equipment should be capable of consistent accuracy of 0.00025 mm.

At this order of accuracy, factors such as temperature and distortion, which normally need not be considered, can become significant.

13.4 EFFECT OF TEMPERATURE

Of such factors, temperature is the most prominent, and although its effect must be recognised, it should not be exaggerated. In a large organisation undertaking a great deal of precision measurement, a standards room or metrology laboratory maintained at a constant temperature should be provided. If this is the case, the standard temperature, which was established internationally in 1932, is 20°C. The provision of such facilities is expensive and a realistic assessment of the requirements of the product may very well reveal that such a refinement is not really necessary.

Changes in temperature will cause all components in an assembled product to expand or contract. They will also cause a similar variations in the equipment used to measure them. What is important, therefore, is not the absolute variation but the comparative variations of one part and another, and of each part and the equipment used to measure it.

Assuming that the product is made of metallic materials, the maximum variations in the coefficients of expansion are likely to be between that of steel, which is around $12 \times 10^{-6}/°C$ and that of light alloys which is of the order of $24 \times 10^{-6}/°C$. If, for example, a dimension of 40 cm is considered, the difference between the two materials over the comparatively wide temperature variation of 10 Centigrade degrees would be $40 \times 12 \times 10^{-6}$ which is 0.0048 mm. The significance of a variation of this order on the function of the parts in the product must then be considered in order to determine whether a temperature controlled measuring environment is likely to be of value.

13.5 EFFECT OF DISTORTION

Distortion as referred to in this context is, in all but extreme precision laboratory applications, largely academic. As an example, if the length of a long bar is to be measured, when the bar is supported horizontally it will distort under its own weight to the, very much exaggerated, shape shown in Fig. 13.1. This creates two problems, first, the horizontal distance between the ends will be very slightly less than the total length of the bar, and, second, the end faces might not be precisely parallel. These problems can be minimised if the bar is supported at points 0.2232 of the length of the bar from each end. These positions are kown as Airy points.

If the bar is to be measured in the vertical position, distortion will

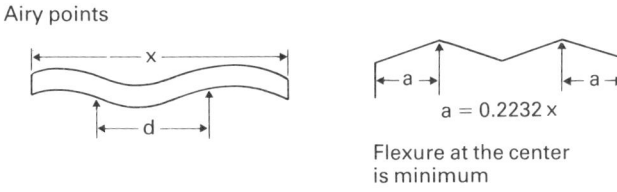

FIGURE 13.1

be due to compression caused by the weight of the bar itself. However, the effect of this on the length would reduce a bar 1 metre long by 0.2 μm and such a small deviation is rarely likely to be significant.

13.6 BASIC REQUIREMENTS

Having determined whether the standards room facility is to be given a temperature controlled environment or not, the two necessities for carrying out the work are basic measurement standards, and means of measurement which are sufficiently accurate to compare shop floor measuring equipment with these standards.

13.7 GAUGE BLOCKS

The universally used basic measuring standards for length are gauge blocks. These were introduced briefly in Chapter 1. They are simply rectangular blocks of metal, generally hardened steel, which are manufactured and calibrated to a very high order of accuracy across their end faces. They are also sometimes known as "slip gauges".

A few gauge blocks are shown in Fig. 13.2, but they are generally supplied in sets, as shown in Fig. 13.3, selected to enable the building up of any length between 0.5 mm and about 20 cm in steps of 0.0005 mm, in the case of metric sizes, or between 0.05 in and about 8 in in steps of 0.0001 in, in the case of Imperial sizes. The upper limit is restricted by the difficulty of handling more than about five blocks.

A very important feature of gauge blocks is the quality of the end faces. The surface finish is better than 0.1 μm and the flatness is better than 0.25 μm. This quality of finish permits a phenomenon known as "wringing". Wringing is achieved by sliding the end faces

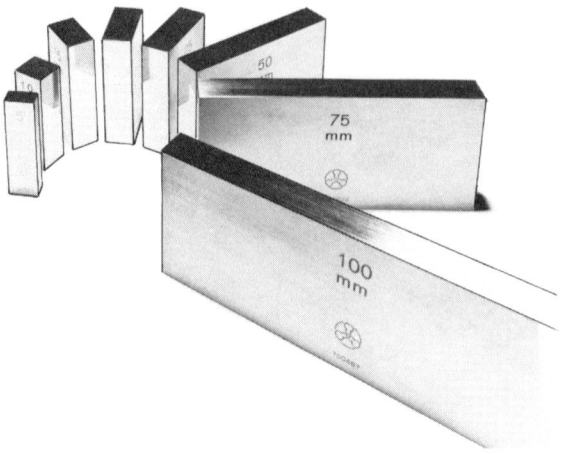

FIGURE 13.2

FIGURE 13.3

of two gauge blocks against each other, usually by a rotating motion, as shown in Fig. 13.4.

This action removes the air film from between the blocks and air pressure then holds the blocks together with remarkable rigidity.

Gauge Blocks 193

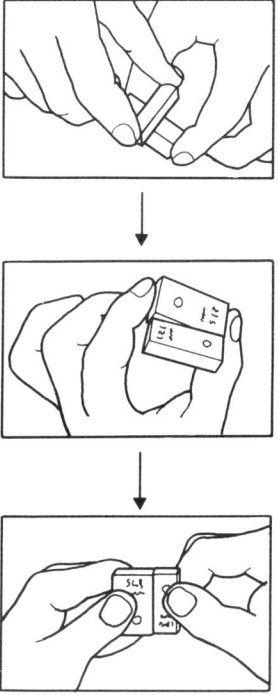

FIGURE 13.4

Under these conditions, the gap between them, even at the level of tolerance looked for in gauge blocks, is effectively nil.

The sizes in a set of 88 metric, or 81 Imperial, are such that the size required can be obtained by wringing together not more than five blocks. The method of selection of the blocks is to consider each decimal place in turn, commencing with the smallest. To build up a dimension of 35.7485 mm, for example, the first block to be selected would be 1.0005 mm. This would then leave 34.748 mm. The second block to select would be 1.008 mm and this leaves 33.74 mm. For the remaining blocks a number of alternatives are possible, but the third block could be 1.24 mm, the fourth 2.5 mm and the fifth 30 mm.

The total is then:

    1.0005 mm
    1.008 mm
    1.24   mm
    2.5   mm
   30   mm
  ―――――――
   35.7485 mm

Gauge blocks are normally made of steel and if steel blocks are used for calibrating a piece of equipment made primarily of steel, the work can be carried out at any atmospheric temperature. Care must be taken, however, to ensure that sufficient time is allowed for all the elements involved to stabilise at the ambient temperature. The time required can be much longer than might be expected and must generally be measured in hours rather than minutes.

Gauge blocks are the means of relating the measurements to some basic external standard. They are normally supplied with a calibration certificate from the manufacturer, but it is possible to obtain, at extra cost, a certificate which relates them to a national standard. The cost of such a certificate is quite small and it is likely to prove to be of good value, not only for the confidence it will give in accuracy of the standards, but also for the impression it will create both within the organisation concerned and with its customers.

Gauge blocks are available in several grades, and those in general use are known as 0, 1 and 2 in metric sizes, and calibration, inspection and workshop in Imperial sizes. There is quite a large difference in cost between the higher and the lower grades and it would be normal to have only one set of 0 or calibration grade and reserve this for use only for calibration purposes.

In general use, gauge blocks can be assembled to give a required size and this is compared with the gauge to be checked. If the gauge is of simple geometry, the comparison can be made using an indicator mounted on a stand, as described in Chapter 4. For standards room work dial indicators as used for shop floor component inspection are unlikely to be sufficiently accurate, and more sensitive devices, which might be mechanical or electronic, are available which are more appropriate to these applications. Two such instruments are shown in Fig. 13.5.

As an example of their use, if it is required to check the diameter of a plug gauge, the method is to set up gauge blocks to the required size, set the indicator at zero with the gauge blocks in position, remove the gauge blocks and roll the plug gauge under the indicator. The indicator will then show the deviation from the required size. When checking a cylindrical gauge, it is important to check for irregular wear by carrying out the diameter comparison at several different positions.

Gauge blocks may be used for many other precision measurement applications with the aid of a variety of accessories. The principle of all these accessories is that, having built up the required dimension with the blocks, end fittings are added and the whole is clamped together in a cage. The accessories consist of extension jaws,

Gauge Blocks 195

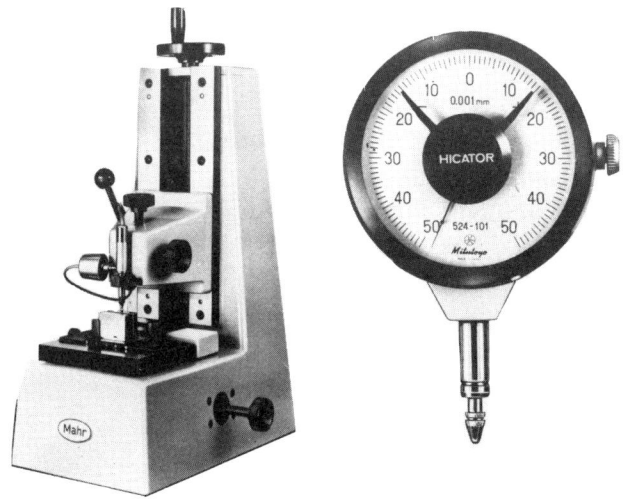

Figure 13.5

scribing points and base blocks and a set, together with the holding cages, is shown in Fig. 13.6. Assemblies of gauge blocks and accessories are shown in Fig. 13.7.

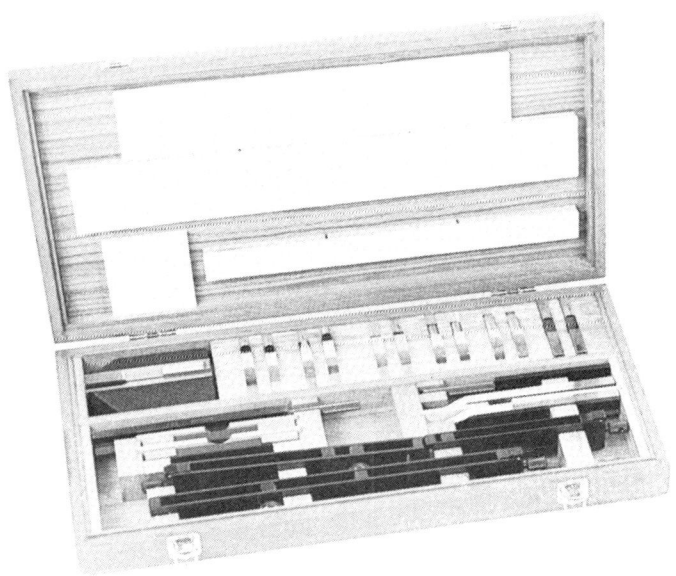

Figure 13.6

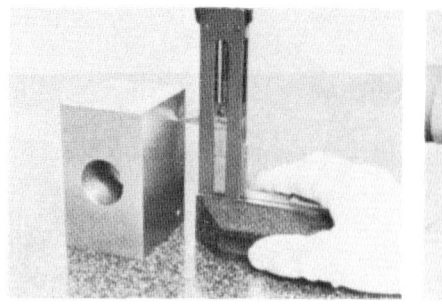
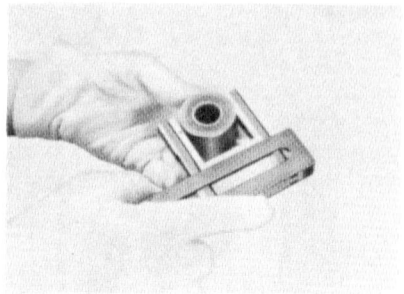

FIGURE 13.7

13.8 END BARS

If it is required to build up a large dimension, beyond the range of gauge blocks, end bars may be used. Although the highest grades of end bars have plain ends and are wrung together in the same way as gauge blocks, for normal inspection this method does not give sufficient stability, and they are usually joined by the less accurate means of a screwed connector. A set of end bars, with accessories, and the detail of the screwed connector are shown in Fig. 13.8.

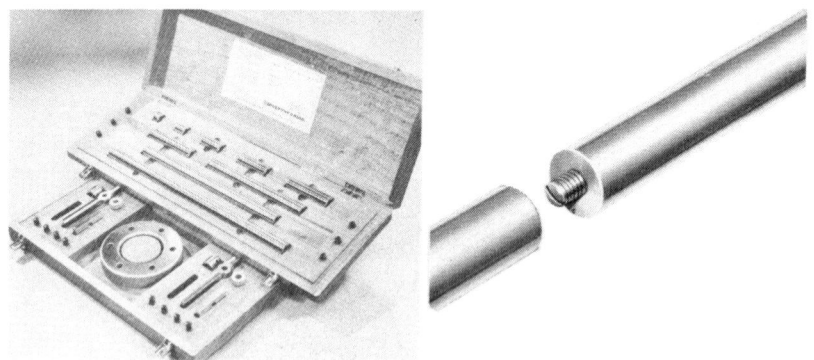

FIGURE 13.8

In order to build up large dimensions in small increments, end bars are used in combination with gauge blocks, as shown in Fig. 13.9. Since the gauge blocks are clamped between the bars and the end accessory, pairs of gauge block combinations are required.

13.9 TOOLMAKER'S MICROSCOPE

Since all instruments capable of measuring to the accuracy required for standards room work are likely to be expensive, it is

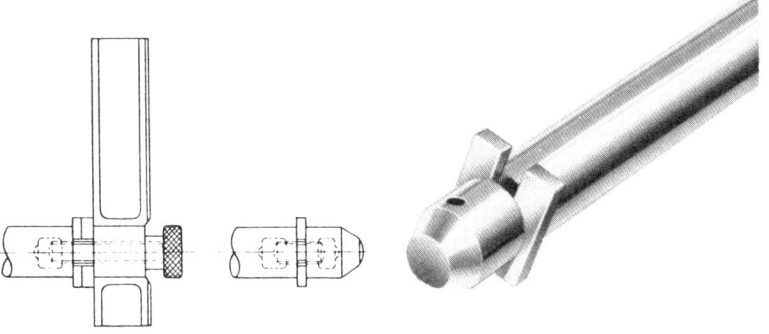

FIGURE 13.9

desirable that they are as versatile as possible. A number of measuring machines satisfying this objective are available, all of which are versions of, or developed from, the toolmaker's microscope. This instrument uses the principles of the optical projector, described in Chapter 11, in that one extreme of the dimension to be measured is aligned to a datum using a microscope, and the gauge is moved so that the other extreme of the dimension is aligned to the same datum. This movement is controlled by accurate measuring scales and the readings obtained against these scales in the two positions enable the required dimension to be calculated.

A simple toolmaker's microscope is shown in Fig. 13.10. It includes a worktable which can be moved in two directions at right

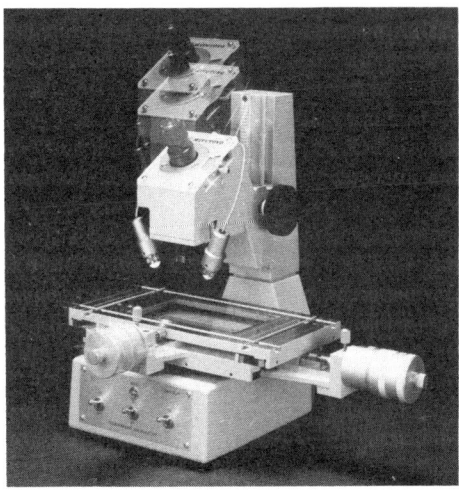

FIGURE 13.10

angles by means of large diameter micrometer drums. The worktable has a transparent centre section so the work being measured can be illuminated in silhouette from below. The microscope itself can be moved vertically on a rigid pillar and attached to it are two small lamps for illuminating the surface of the work.

In spite of its simplicity, an instrument such as this is capable of undertaking a very large variety of precision measurement tasks.

A feature of the toolmaker's microscope is the facility for using interchangeable reticles against which the workpiece is viewed. If the microscope is to be used as a means of locating a feature of the workpiece against a datum, reticles of the type shown in Fig. 13.11 might be used.

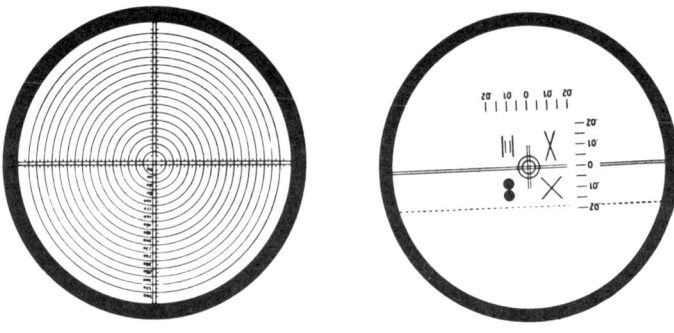

FIGURE 13.11

The microscope might, however, be used for locating a thread for measurement of its position relative to other features, or for examining details of the thread itself. For such purposes as this, reticles as the example in Fig. 13.12 are available.

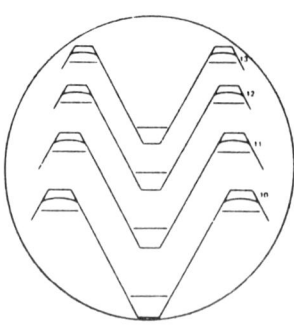

FIGURE 13.12

As an indication of the scope for elaboration on the toolmaker's microscope theme, Fig. 13.13 shows a larger instrument which includes not only a binocular microscope, but also a TV camera for displaying the image on a screen. This instrument has digital readout of the table position in addition to micrometer drums.

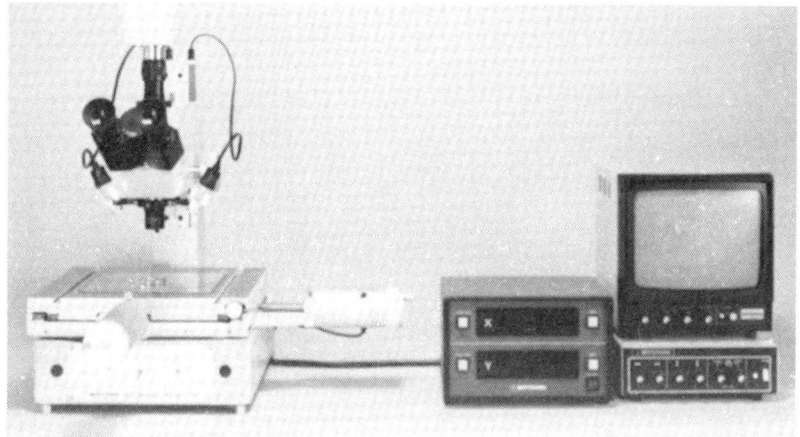

Figure 13.13

Figure 13.14 shows another elaborate version of this type of instrument. This machine features optical micrometer reading of the table position which, together with a very rigid structure, enables measurements to be taken with consistent accuracy to around 1.25 μm (0.00005 in).

Figure 13.14

200 Standards Room

The view through the measuring microscope, at a dimension of 0.6116 mm is shown in Fig. 13.15.

A more recent development of this instrument, shown in Fig. 13.16, gives prominence to the use of a probe, rather than a microscope, for location, although a microscope is also provided as an alternative.

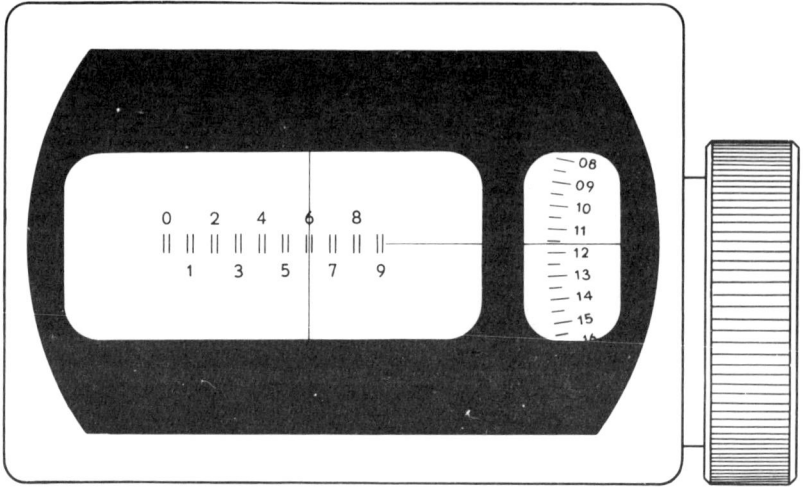

FIGURE 13.15

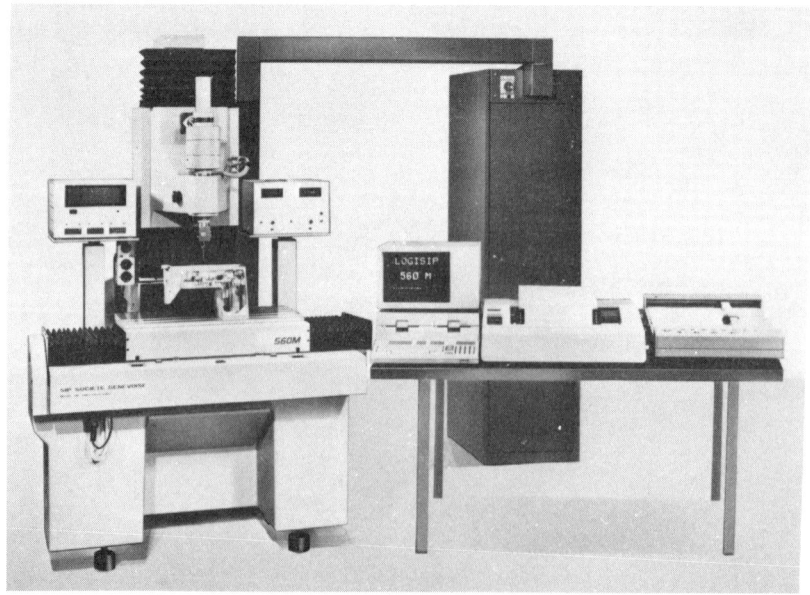

FIGURE 13.16

The resemblance between this machine and the multi-axis machines in Chapter 5 will be seen, suggesting that the toolmaker's microscope has now become a high precision co-ordinate measuring machine.

Machines of such size and versatility, with the capability of standards room accuracy, are inevitably very expensive. For this reason, if for no other, the temptation to obtain and use such machines for work more appropriate to the machines described in Chapter 5 must be resisted.

However, a mid-range toolmaker's microscope machine, together with a granite surface table, gauge blocks and accessories, and accurate comparators on suitable stands are likely to be able to cope with the majority of standards room tasks. One other instrument which could be valuable in this environment is a surface finish measuring device. In this situation, it would be used for establishing shop floor standards and possibly calibrating samples for production use. One of the more comprehensive of the machines described in Chapter 7 would be suitable.

13.10 SPECIAL PURPOSE MEASURING MACHINES

In the context of a chapter so far concerned with the special requirements of gauge checking, it is perhaps appropriate to consider measuring equipment devoted to other specific tasks.

Whilst reference was made in Chapter 5 to multi-axis machines which might be tailored in their overall capacity for specific purposes, it can be economical to provide machines which are designed individually to measure parts of a particular type, or even one specific part.

Again, mention was made of such machines in connection with the measurement of blades in Chapter 9, but the principle can be used for other components. The machine shown in Fig. 13.17 for checking small cylindrical components is an example.

This machine is capable of checking up to 20 dimensions, as the example in Fig. 13.17, in one operation.

The criterion for the justification of such equipment is, of course, largely economic. It is generally expensive and limited in its applicability. On the other hand, it is likely to be very quick, consistently accurate and require less skill to operate than normal general purpose equipment.

It is necessary, therefore, to determine whether the complexity of the task and the quantities to be inspected are such that the value of the time saved and the lower level of skill required will, over a

FIGURE 13.17

specified period, cover the additional cost of the equipment. An excellent example of an obviously viable use of a machine in this category is the blade airfoil inspection machine described in Chapter 9, and other examples of machines of this type are mentioned in connection with in-process inspection in Chapter 14.

A fairly recent development in this field is the inspection machine which is supplied as a kit of parts from which can be built virtually an infinite variety of single purpose machines within the overall limitations of the kit. Such a kit of parts is shown in Fig. 13.18.

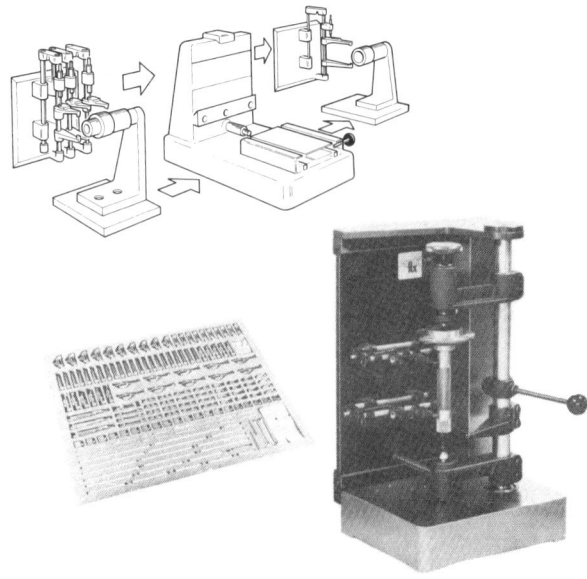

FIGURE 13.18

Special Purpose Measuring Machines 203

It consists of two groups. The first is essentially a base plate containing locating fixtures for the part to be measured and for the measuring probes, and the measuring probes themselves. The second group contains the indicating devices, which can include specific size recorders or acceptance limit indicators for either individual dimensions or the part as a whole.

All special purpose machines, whether designed for a unique application or built up from kits, are likely to function as comparison devices and will give more accurate results when used in that mode. It is normal for the machines to be set against a master, which can be either a specially made simulation of the part to be inspected, or a sample of the actual part. A part used for this purpose need not be precisely correct to the specification, indeed a scrapped part can often be satisfactorily used. What is important is that the relevant dimensions are known to a degree of accuracy of about 10% of the tolerance on the production part. This is likely to involve measuring the sample in the standards room, on a machine of the toolmaker's microscope type.

The practicability of using an actual part as a sample will depend to some extent on the material from which it is made. It is important that it is sufficiently hard to be inserted into the measuring machine a large number of times without showing any discernible wear. If the normal production material will not stand up to this, then a sample must be produced from a harder material. Although the production drawing can be used as the basis for this sample, it is not necessary for it to include intricacies which are not relevant to the dimensions being measured. Figure 13.19 indicates how a sample, or master might relate to a production part.

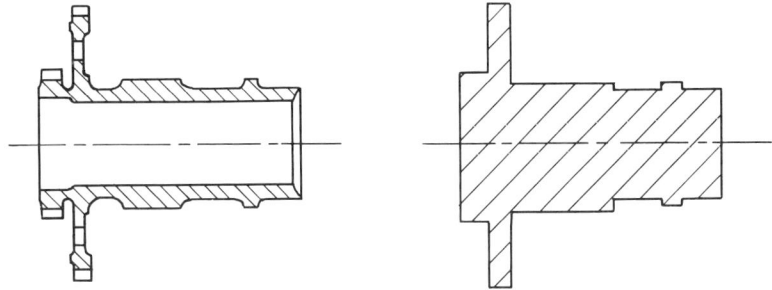

FIGURE 13.19

In normal use, it is important that the measuring machine is reset against the sample at regular intervals. These can be either after a given number of parts have been checked, or at specified time intervals, say the beginning of each shift, or a combination of both.

Chapter 14

In-process and In-cycle Measurement

The equipment described so far in this book has been aimed at inspecting a part, dimension or characteristic after it has been produced and removed from the machine which produced it.

If it can be done practically and economically, it is preferable to carry out the inspection during the manufacturing cycle. In this event, it might be possible to make sure that a particular dimension cannot be incorrect, or, at least, make sure that further work is not wasted on a part already likely to be scrap because of an undetected incorrect dimension.

A certain amount of equipment is available which makes inspection at this stage possible, and it can be done in two basic ways.

The terms "in-process" and "in-cycle" gauging can be used to mean the same thing, but it is convenient to define them to refer to alternatives.

In-process gauging will be used to refer to measurement carried out whilst the process for producing the dimension concerned is still in operation. In-cycle measurement can then refer to measurements taken after the dimension concerned has been produced, but before the part has been removed from the machine. This category can also include work done by measuring devices which are contained in a group of machines through which the part passes in sequence.

14.1 IN-PROCESS GAUGING

The simplest means of measurement whilst the operation is in progress is to make use of the measuring system built into the machine tool itself. Unfortunately, however, such systems, although adequate for making small movements of the cutting tool, are rarely accurate enough for absolute, total measurement.

A way of overcoming this is to add an additional, and independent measuring system.

Figure 14.1 shows such a system, and this consists of non-magnetic steel rails rigidly mounted on the moving bed of the machine, to which are attached magnetic mild steel "targets" which pass before electromagnetic sensors rigidly attached to the base of the machine. The output from the sensors varies sufficiently with small movements of the target to be able to detect movements as small as 0.0002 in.

FIGURE 14.1

The equipment is used comparatively in that the targets are set whilst machining a sample component which is subsequently measured and suitable corrections applied. For follow-on quantities, the indicator to which the output from the sensors is fed is used to determine when the cutting tool has produced the correct size.

A rather different system is illustrated in Fig. 14.2.

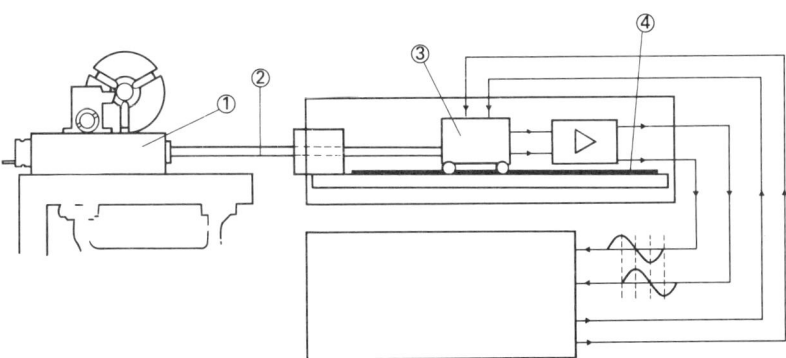

FIGURE 14.2

206 In-processing and In-cycle Measurement

In this system, the cross-slide (1) carrying the cutting tool is linked by a push rod (2) to a reading head (3) which moves across a scale (4). The measuring device is optical, similar to that described in Chapter 5, and its output is fed to a display unit mounted on the machine. Again, the principle is to assess the size of the component by measuring the movement of the cutting tool.

Available equipment capable of carrying out measurement directly on the part whilst the dimension being measured is still being produced is largely confined to outside diameters on fairly simple components. An example of a typical device for measurement during a cylindrical grinding operation is shown in Fig. 14.3.

This device is essentially an adjustable caliper with one fixed anvil and one moving anvil spring loaded against the workpiece.

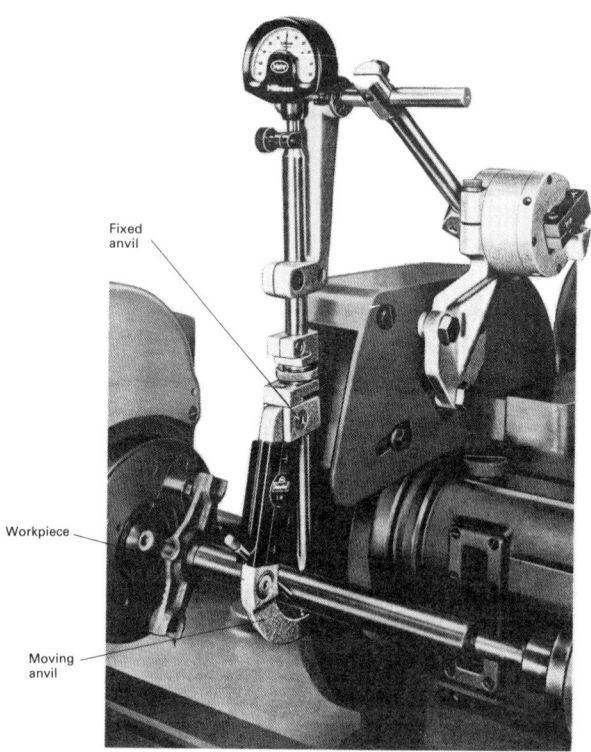

FIGURE 14.3

The whole is also spring loaded so that the fixed anvil contacts the workpiece, and a centre stop is provided which will locate the fixed and moveable anvils across a diameter when it has reached the desired size.

The device is, of course, used as a comparator and it has to be set against a master close to the required size of the work to be measured. Its use in this way automatically compensates for wear which inevitably takes place through contact between the anvils and the rotating work, although this is minimised by using very hard materials, such as diamond and tungsten carbide, for the contacting surfaces. It is also important that the device is robust enough to be unaffected by cutting lubricants, metal swarf and grinding dust.

This particular piece of equipment merely records the size of the diameter at that particular time, and it is necessary for the machine operator to observe the reading continuously and to take action accordingly. It is a fairly obvious step to record the size as an electrical output and to make use of this to exercise some direct control over the machine tool.

The instrument shown in Fig. 14.4 is considerably more elaborate in several respects.

The measuring head is mounted on a hydraulically powered base which moves it into or away from the measuring position.

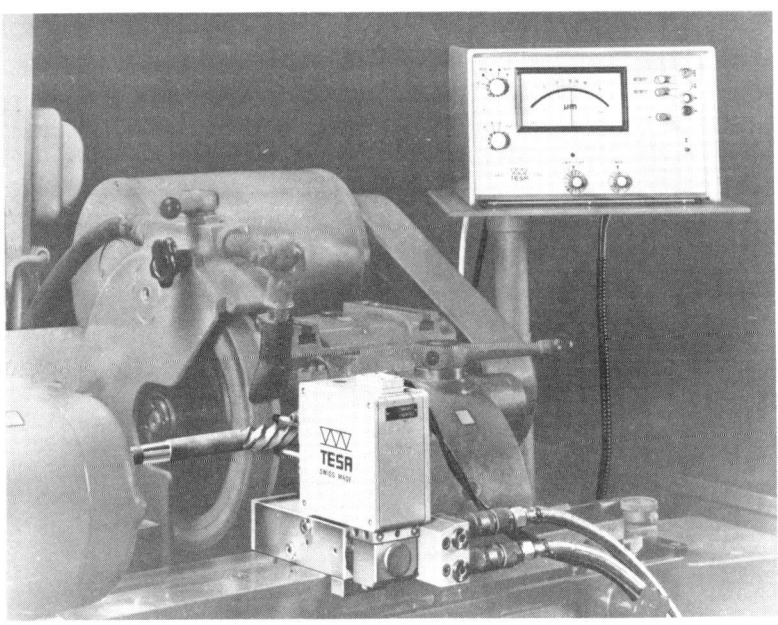

FIGURE 14.4

In addition to operating an analogue indicator, the output from the measuring head can be used to control the rate of feed on the machine tool, in such a way that it reduces as the part approaches the required size.

208 In-process and In-cycle Measurement

The diagrams in Fig. 14.5 indicate the various stages in the total cycle, each of which is indicated by illuminated lights on the display unit.

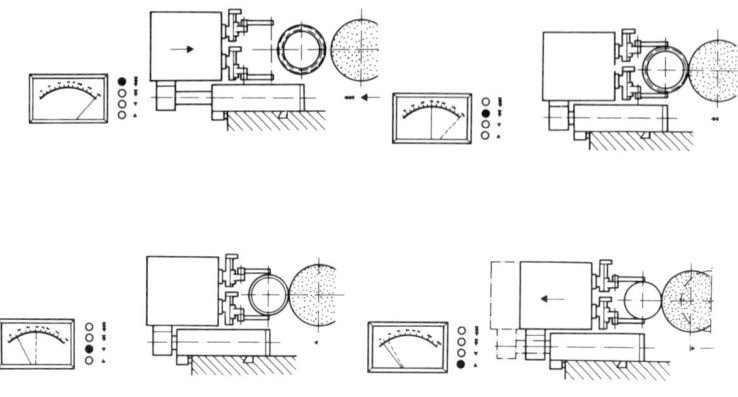

FIGURE 14.5

With the addition of suitable accessories, equipment of this type can be used to control a variety of machining operations. Surface grinding and internal grinding are shown in Fig. 14.6.

FIGURE 14.6

14.2 IN-CYCLE GAUGING

A type of control which does not check whilst the machining operation is actually in progress is illustrated diagrammatically in Fig. 14.7. This layout shows a cutting tool (1) boring a component (2). An electrical bore gauge is mounted at (3) and the machine is programmed to move the component automatically onto the gauge.

In-cycle Gauging

The signal from the gauge is transmitted to amplifying equipment (4) and (5) which compares the size recorded with the size required. If adjustment is necessary, an electrical signal is sent to a motor (6) which, via a drawbar (7) adjusts the cutting tool.

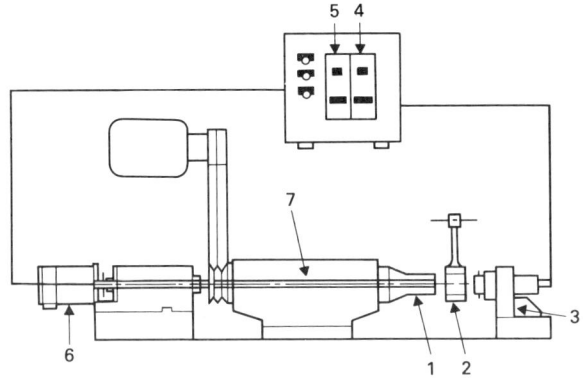

FIGURE 14.7

The devices shown in Fig. 14.3. and in Fig. 14.4 actually measure the work as it is being cut, whilst the arrangement of Fig. 14.7 moves the part from the cutting tool and presents it to a gauge. For high precision the second method is preferable because the size will be affected by the heat generated during the cutting process. On the other hand, the gauging process takes additional time and the total cycle will be slower.

Another means of in-process control of this second type makes use of a probe of the type shown in Fig. 14.8

This probe is a development for use on machine tools of the touch trigger probe described in Chapter 5.

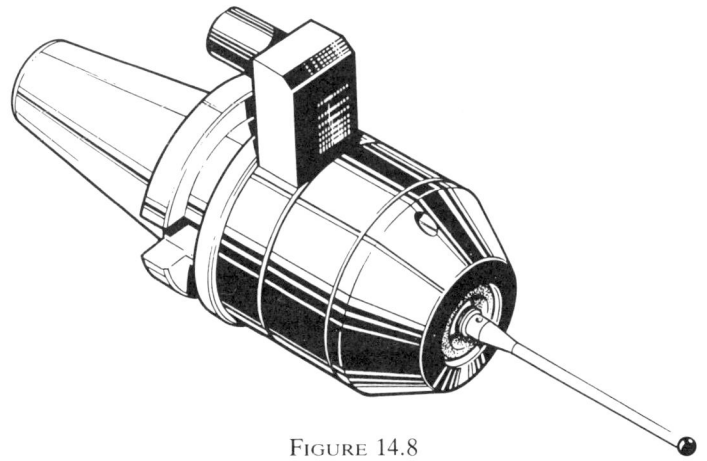

FIGURE 14.8

210 In-process and In-cycle Measurement

The basic technique for using such a probe is to mount it so that it can replace the cutting tool in the machine, as shown in Fig. 14.8. It can, in fact, be located in one of the interchangeable tool locations in the tool turret of a machining centre, as the two smaller probes in Fig. 14.9.

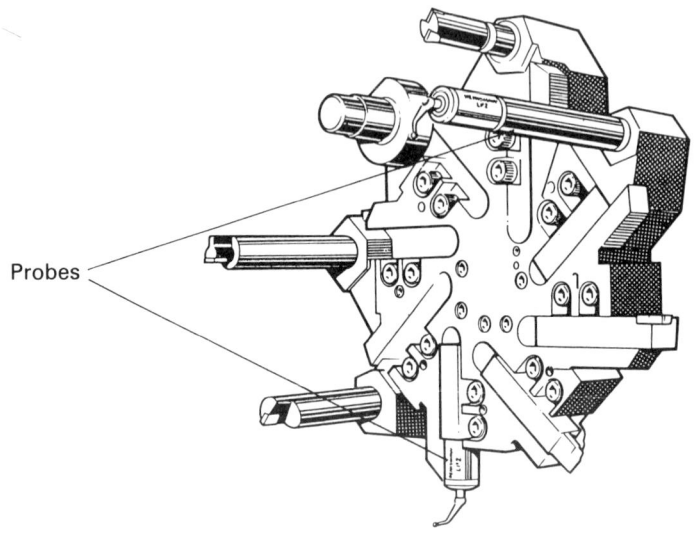

FIGURE 14.9

If a probe is to be used in this way, it is obvious that the means of transmitting the signal from the probe to the readout device or processing system must be able to operate whilst the probe is rotating, and it must also be capable of easy disconnection.

Two methods of achieving this are available. The first is an inductive transmission system shown diagrammatically in Fig. 14.10.

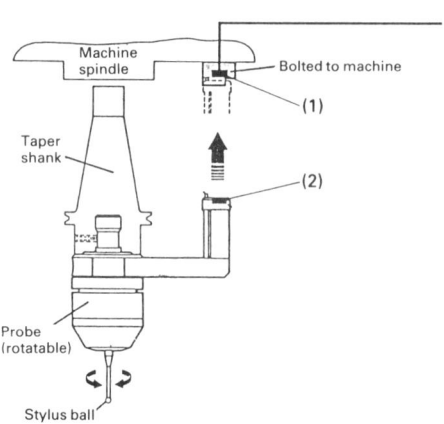

FIGURE 14.10

In-cycle Gauging

The receiver of the inductive coupling (1) is permanently attached to the machine tool, whilst the transmitter (2) is located at the end of an arm which forms part of the probe housing. The probe itself is free to rotate relative to the arm, which is static when the probe is in position. The coupling operates with a gap of up to 2 mm between its two elements.

The second system is optical, as illustrated in Fig. 14.11.

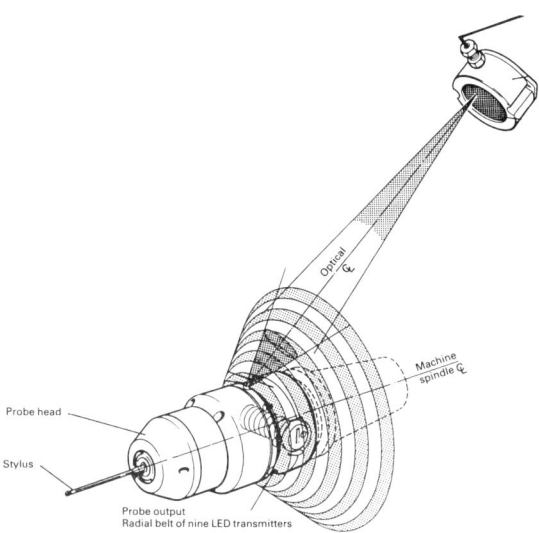

FIGURE 14.11

This particular system has a ring of nine infrared transmitters which produce a continuous 360° signal as the probe and its housing rotate. This signal is picked up by a receiver which may be up to 3 m away. The readings may be displayed on a digital readout device or used to trigger changes in the machine tool settings.

Whilst in-process measurement might be ideal in some respects, it does suffer from some disadvantages. It has already been noted that dimensions can be affected by the heat of the machining operation, and some distortion might also occur due to clamping of the part in the machine. This is quite common and important to avoid, but it will not be detected if measurement takes place with the part in the distorted state.

If the measuring procedure makes use of the measuring system within the machine tool, the accuracy of this system should be checked. It is possible to overcome these inaccuracies to some extent by using a component checked by other means as a comparison.

212 In-process and In-cycle Measurement

Another argument against the form of measurement which takes place whilst the part is still in the machine tool is that, whilst the measurement is being made, the machine is not performing its primary function and the result can be low machine utilisation. This is particularly important if the machine tool is expensive.

The other form of in-cycle gauging, which overcomes this point, can, in the extreme case, be a normal piece of inspection equipment but dedicated to one particular activity. To qualify for the "in-cycle" description, it would normally be part of a collection of processing machines geographically situated so that the complete manufacturing cycle of a specific part is achieved by moving it sequentially between machines which are close, if not adjacent, to each other.

Figure 14.12 shows a production line layout in which the product moves between adjacent machines along a moving platform.

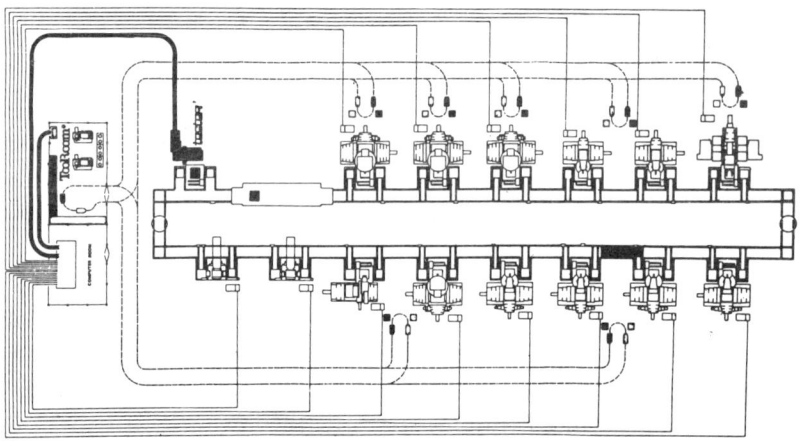

FIGURE 14.12

The layout includes an automatic inspection station, shown in Fig. 14.13, which is linked to a tool room controlling the various machine tool settings. This enables the appropriate manufacturing parameter to be automatically corrected if the inspection machine reveals either a defective feature or a trend towards rejection. Such an arrangement is, of course, only likely to be justified if quantities are very large, sufficient, in fact, to be made continuously by the machine tool complex concerned. It does, however, lend itself to 100% inspection, i.e. complete inspection of every part, and to inspection of the part at stages during manufacture. This ensures that wasted effort is kept to a minimum and that the quality of the finished parts is very high. It is the reason why, in terms of

In-cycle Gauging 213

FIGURE 14.13

conformance to specification, mass produced parts are generally of higher quality than parts produced individually.

Mass production lines are a particularly suitable environment for special purpose machines such as the machine shown in Chapter 13, Fig. 13.17. In these situations the machines are likely to have parts automatically fed to them. Figure 14.14 shows a machine for checking the cylinder bores of a crankcase which is fed by a mass production transfer line.

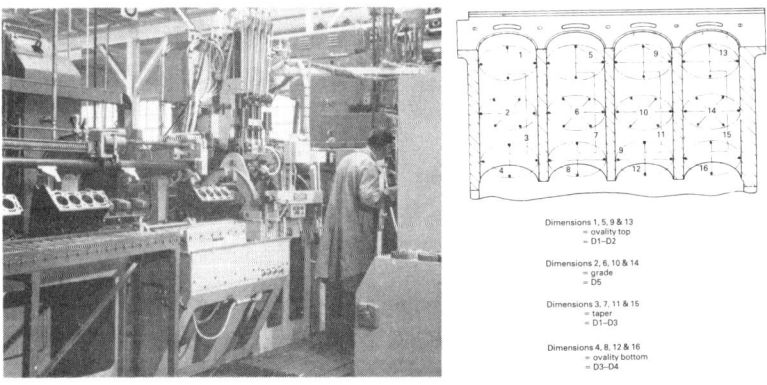

FIGURE 14.14

This machine checks and processes the 16 dimensions shown in about 1 minute. The gauging system is pneumatic but readings are converted to electronic signals for processing.

214 In-process and In-cycle Measurement

If the inspection machine is actually in the production line, it is possible for it to do a number of things automatically if any feature of the part is not correct. It will certainly indicate by visual or audible signal or both, that it has detected an error. It is likely to be able to indicate by visual signal which of the features checked is in error, and by how much. It is also possible for it to stop the flow of parts, to correct the manufacturing machine responsible automatically, and to stamp or paint an identification mark on the defective part, possibly adjacent to the defective feature.

Another means of feeding inspection machines in quantity is shown in Fig. 14.15.

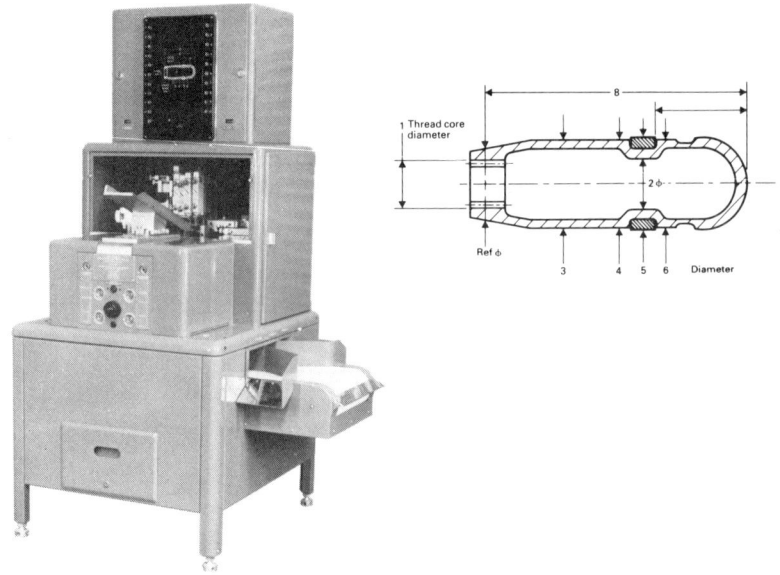

FIGURE 14.15

This machine is magazine loaded and checks the 8 dimensions shown at a rate of 900 per hour. After inspection, the parts are deposited on a moving belt which segregates them into three groups, correct, incorrect but correctable and incorrect but not correctable. The output chutes may be seen on the right-hand side of the machine.

An inspection machine which carries out a number of checks on length, diameter, concentricity and alignment on a small shaft is shown in Fig. 14.16.

This machine is fed from a vibrating bowl which causes the parts to be fed continuously to the measuring station. The machine will also segregate the parts into groups after inspection.

In-cycle Gauging 215

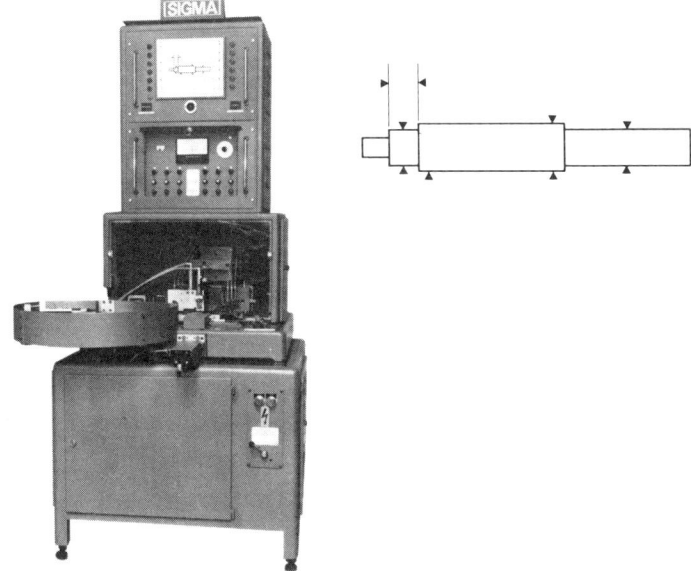

FIGURE 14.16

As mentioned in Chapter 3, it is occasionally necessary to assemble parts with such a close fit that tolerance to achieve this automatically would be too tight for economical manufacture. In such cases, relevant dimensions of the mating parts have to be measured and graded.

An extension of the machines in Figs. 14.15 and 14.16, which is designed specifically for the purpose of grading is shown in Fig. 14.17.

FIGURE 14.17

216 In-process and In-cycle Measurement

This machine checks five features on automotive wrist pins, including rotating them to check roundness and sorting them into four acceptable categories for this feature. In addition, the machine segregates oversize and undersize and marks the pins with a grading symbol. This machine is able to check up to 4000 parts per hour.

It is possible to combine the various features of the machines referred to in this chapter, and Fig. 14.18 shows a large machine for incorporation into a production line which is capable of accepting, rejecting and grading at high speed a variety of small precision parts.

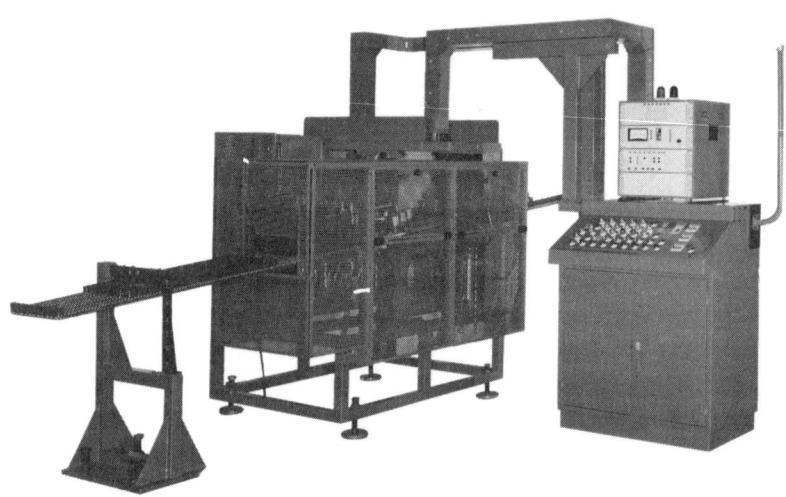

FIGURE 14.18

Postscript

From the progression through just a selection of the range of metrology equipment available, from steel rule to fully automatic, high precision machine with computer operation, feedback and control, it is clear that there is much more to come.

In the near future there will be many other tempting examples of highly sophisticated devices, exploiting the enormous potential of computers, lasers and, no doubt, new technical discoveries not yet even imagined.

It is possible that the ownership of such marvels will, in itself, be considered sufficiently prestigious to justify their purchase on these grounds alone. Otherwise, however, the message of this book will remain. The most efficient methods will be those which ensure that the product is within the requirements of its specification, without either the capability of unnecessarily high accuracy, or expensive and elaborate additional features never likely to be required.

A product is invariably purchased to conform to a specification at a price. The specification is likely to imply, if not specifically state, a standard of performance, a standard of reliability, an expectation of life and, possibly, a standard of appearance. If all the specification parameters are met, the quality of the product cannot be improved. The science necessary to make quantified judgements of these parameters is metrology.

Index

Abbe's Law 13–14
Accuracy 12–13, 15, 189
Air bearings 64
Air gauging 155
Airy points 190
Alignment telescope 158
Analogue column indicators 130
Analogue displays 43
Analytical statistics 67
Angle 74
  Dekkor 163
  plates 35
Autocollimator 162
Axis of measurement 13

BA threads 134
Backlash 117
Ball-ended probe 68
Base circle 104, 106–107
Bench micrometer 49
Bevel gears 102
Bevel protractor 77
Bilateral 18
  tolerance 17, 19
Blade airfoil inspection 202
Bore gauges 52–53
Bore micrometer 53
Box cubes 35
Bridge 63
  pressure system 45
Bridge-type machine 65
Brooks level comparator 2
BSF threads 134
BSP threads 134
BSW threads 134

Calibration certificate 194
Caliper gauge 14
Caliper-type micrometer 13

Calipers 29, 37
Cantilever type 63
Centre line average (CLA) 91–92, 96
Clinometer 80
Co-ordinate measuring machine 201
Coating of threads 145
Collimators 158, 160–161
Comparison 15
  devices 203
Compound sine table 83
Computers 67, 174–178
Concentricity 26, 83
Concession 8
Conical probe 68
Contact probes 69
Contoured surfaces 121
Crest 132
Cut-off value 93, 97

Datum 26
Degrees 74
Depth gauge 31–32
Dial caliper 37
Dial indicator 68, 78, 113
  lever type 39
Digital 43
  electronic height gauge 60
  electronic micrometers 50
  indicator 130
  readout 13, 38, 51, 86
Display systems 37
Distortion 13–14, 189–190

Eccentricity 87
Eddy current methods 186
Eden-Rolt comparator 3
Effective diameter 132–133, 136–137
Electronic 43
  measuring systems 58
  digital read-out 57
End bars 196

220 Index

Fibre optics 154–155
Finish 91–92
Fits 22
Five-axis machine 65
Flanks 102, 117, 132
Flatness 91, 93, 101, 167–168
Floating carriage micrometer 138
Form 93, 101
Four-axis machine 65
Four-poster multiaxis machine 65
Fringes 59, 168
Function control 174–175

Gamma radiation 186
Gap gauge 147
Gauge blocks 5, 181, 191, 193–194
  accessories 195
Gauge tolerances 146
Gear teeth 102
  measuring machines 182
Geometrical tolerances 24
GO 9, 54–55, 142
GO/NO-GO 10
Goniometric microscope 135
Grading 215
Granite 34, 62
Gratings 59–60

Height gauges 31, 33, 36, 17, 10, 43, 173
Height micrometer 52
Helical gears 102
Helix angle 109
High definition radiography 183
Hole 20–22

Illumination 149
Imperial dimensions 39
  standard yard 1–2
In-cycle 204, 208, 212
In-process 204, 211
  gauging 70
Inductive transmission 210
Inspection station
  automatic 212
Interference 22
  fringes 167

Interferometry 165, 169
Internal micrometer 52
Involute 103
ISO (International Standards Organisation) system 24

Johansson 5

Laser 170, 173
  light 170
Lay 94
Lighting intensity 149
Linear conversion 76
Linear measurement 29
Liquid manometer 45
Lobing 53

Magnetic system 58
Magnetic bases 43
Magnification 12–13, 150, 153
Major diameter 132, 136, 142
Mass production lines 213
Master 203
Maximum metal condition 18
Measuring probes 60, 62, 68
Mechanical measuring instruments 29
Metallurgical procedures 174
Metre 1–3
Metre des archives 1–2
Metrology laboratory 185, 190
Micro-inches 94
Microfocus X-ray 183
Micrometers 4–5, 13–14, 46, 48–50, 52–54, 138, 147, 159–60
  anvils 168
  sets 48
Micron 94
Minor diameter 132, 142
Miximum material condition 17
Monochromatic light 166, 168
Motorised probe head 72
Multi-axis measuring machines 6, 10–11, 63, 65, 128
Multi-probe measurement 128

NO-GO 9, 54–55, 142
Non-contact 45

Index 221

gauging 170
measurement 155
probes 155, 170
Non-destructive testing 183
Non-working flanks 117

Optical flat 166, 168
Optical micrometer 159–160
Optical parallels 168
Optical projection 121, 151
Optical vernier 85
Output processing 176
Ovality 53

Parallax 130–135
Parallelism 24–25, 168
Peak to mean 91
Peak to valley 91
Photoelectric system 164
Photoelectric autocollimator 59
Pitch 110, 113, 132, 134
circles 104
line clearance 117
Plug gauges 52–53, 142, 194
Pneumatic bases 43
Pneumatic gauge 43–44
Pneumatic indicators 43
Pressure angle 104, 111
Primary texture 93
Principle of comparison 15, 41
Process capability 27–28
Projecting the cross-section 124
Projection 123, 152, 155
Protractor 76

Quality assurance 7, 9
Quality control 7, 9
Quantified manufacturing tolerances 4
Quantity 11

Radians 74
Radiography 183
Radius gauges 127
Repeatability 12
Replicas 100, 141
Resolution 12–13

Reticles 135, 140, 198
Rigid probes 68
Ring gauges 55, 143–144
Rolling tests 115
Root 132
relief 109
Root mean square 91
Rotating tables 83
Roughness 93
Roundness 24–25, 83, 87, 89

Salvaging 8
Sampling lengths 92–94
Scrap 8
Screw threads 132
Scribing block 42
Secondary texture 93
Selective assembly 20
Shadow 122
Shaft basis 20–21
Shoe 96, 98, 101
Sine bar 81
Sine table 82
Single flank machines 119
Single flank tester 119
Skid 96, 98, 101
Slip gauge 191
Snap gauge 55–56, 143
Sorting
right/wrong type 187
Special purpose machines 203, 213
Specification 8–9
Spin table 87
Spirit level 79
Spur gears 102
Squareness 24, 26
Standard tolerance systems 19
Standards room 185, 190
Steel rule 29
Stick micrometer 49, 54
Stylus 71, 96–98, 101, 104, 108, 126
Surface
finish 91, 201
gauge 42
illumination 153
plates 34
projection 122

table 33, 36
  box cube 33
  cast iron 34

Telescopes 158, 161
Temperature 189–190
Templates 127–128
Threads 132, 134
  coating 145
  tolerances 144
Thread measuring cylinders 136, 138, 148
Three-axis machine 62
Tip relief 109
Tolerance 4, 8, 9, 17, 20, 189
  systems 21
Toolmaker's microscope 135, 140–141, 196–198, 201
Tooth
  caliper 110
  markings 119
  thickness 110–112
Touch trigger probe 70
Transition 22
Two-axis machine 61

Ultrasonics 185
UNC 134
UNF 134
Unilateral tolerance 17–19

Vee blocks 33, 35–36
Vernier caliper 30–32, 37, 111
  height gauge 31–32
Vernier scale 30–31, 47, 77, 80, 84, 86
Video camera 157–158
Virtual effective diameter 141
Visual assessment 149–150

Wavelength
  of light 3
Waviness 93
White light 168
Whitworth 134
Working flanks 117
Worm gears 102
Wringing 191

X axis 62
X-rays 183–184

Y axis 62
Yard 1–2

Z axis 62

Renishaw Probes.
Can you afford CNC without one?

Your competitors can't, and it's hardly surprising, just look at the benefits.

Automated set-up (up to 95% time savings are possible). On-machine dimensional checking and self-correction of errors. Real reductions in operating costs and scrap and there's improved quality control.

If you're looking at CNC to satisfy your production needs, talk to us first. The payback can be faster than you think.

Let us help you, by contacting

RENISHAW

Renishaw Metrology Ltd.,
New Mills, Wotton-under-Edge,
Gloucestershire GL12 8JR.
Tel: (0453) 844211

Quality Assured with Renishaw

Britain's Leading Consultants in Quality Improvement

The symbiotic relationship between Metrology and the Science of Quality is evident. Pressure of competition in industry makes it increasingly necessary for metrology specialists to understand fully the Science of Quality. Conscious of this need David Hutchins Associates, Britain's leading Consultants in Quality Improvement, have devised a comprehensive range of courses on quality-related subjects, available through the DHA Quality College. The courses offered by the DHA Quality College include many of direct relevance to the topics covered in this publication.

In addition to the range of courses available through the DHA Quality College, DHA are able to offer in-house training on quality, tailored to the client's specific requirements. For those clients particularly concerned with metrology and modern inspection techniques DHA offer the expertise of the author of this publication, Derek Anthony, an Associate Consultant with DHA.

DHA are also the sole UK licensors for the internationally renowned Juran Institute and offer the range of Juran educational products and services designed to teach the theory and practice of 'Managing for Quality'.

For further information on any of DHA's consultancy services, external or internal courses and the Juran range of products and services, please contact Frank Glenister, Managing Director, at the address below.

DAVID HUTCHINS ASSOCIATES LTD
Consultants in Quality Improvement
13/14 Hermitage Parade, High Street, Ascot, Berkshire, SL5 7HE. Tel: (0990) 28712

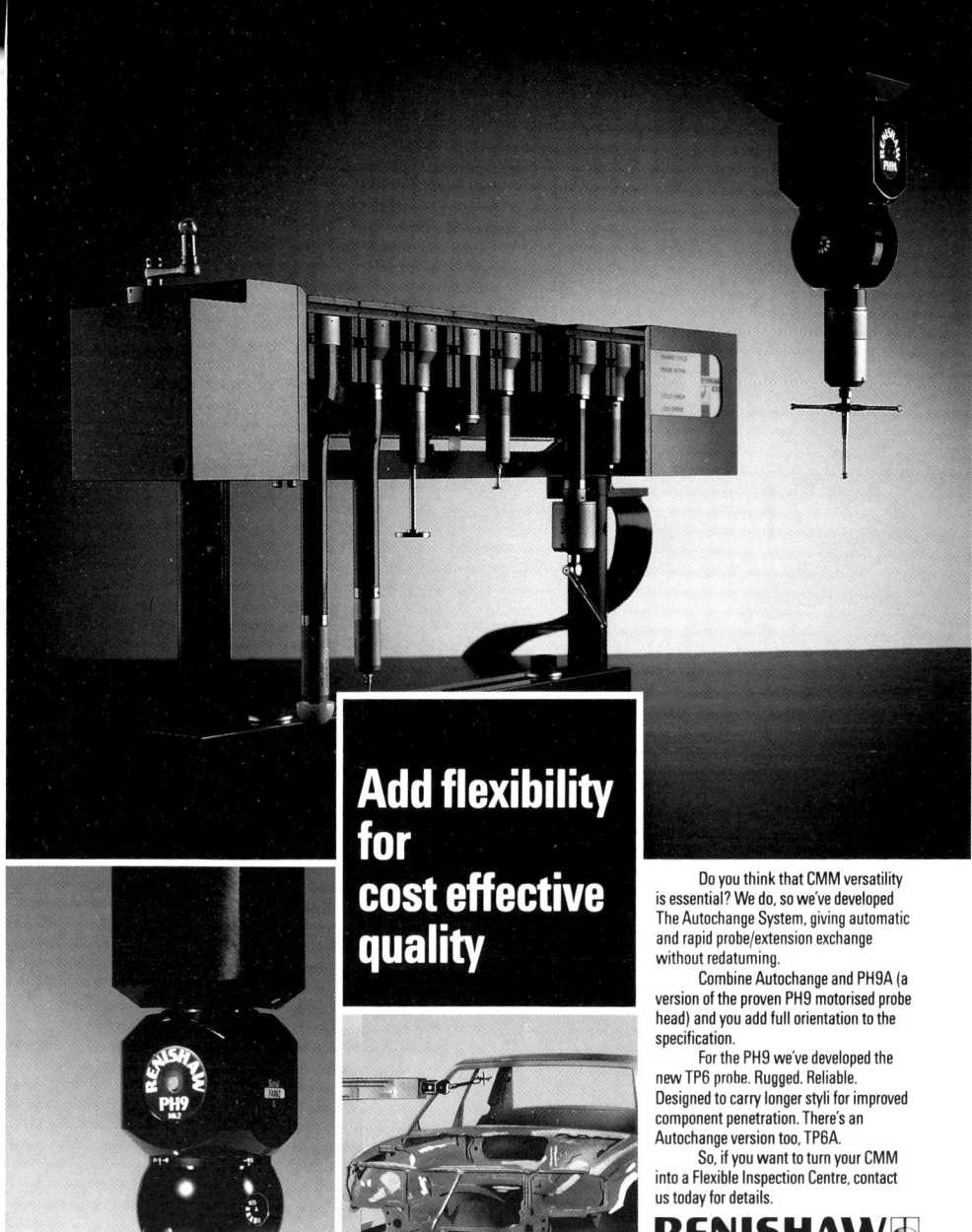

Add flexibility for cost effective quality

Do you think that CMM versatility is essential? We do, so we've developed The Autochange System, giving automatic and rapid probe/extension exchange without redatuming.

Combine Autochange and PH9A (a version of the proven PH9 motorised probe head) and you add full orientation to the specification.

For the PH9 we've developed the new TP6 probe. Rugged. Reliable. Designed to carry longer styli for improved component penetration. There's an Autochange version too, TP6A.

So, if you want to turn your CMM into a Flexible Inspection Centre, contact us today for details.

RENISHAW

Renishaw Metrology Ltd.,
New Mills, Wotton-under-Edge,
Gloucestershire GL12 8JR, United Kingdom.
Tel: (0453) 844211

Quality Assured with Renishaw

Keeping ahead in quality

WEIGHING
NEWS & VIEWS
MEASUREMENT IN ACTION
COMMENT   PROFILE   COMPUTERS
PUBLICATIONS   THERMOMETRY   SERVICES
HARDWARE ... SOFTWARE
Quality Today Interview   TESTING
QUALITY IN PRACTICE
APPOINTMENTS
EQUIPMENT

Industry needs QualityToday

THE MAGAZINE FOR

Measurement & Inspection Technology

For your sample copy
Telephone:
Jean Reed on
(0622) 59841

**Earl House, Earl Street, Maidstone,
Kent ME14 1PE**